The Revd Dr John Polkinghorne, KBE, FRS, is Fellow of Queens' College, Cambridge. Formerly Professor of Mathematical Physics at Cambridge University, he is a priest and Canon Theologian at Liverpool Cathedral. He won the Templeton Prize for Science and Religion in 2002 and is the author of many books, including *Exploring Reality: The intertwining of science and religion* (2005), *Quantum Physics and Theology* (2007), *Theology in the Context of Science* (2008) and an autobiography, *From Physicist to Priest* (2007).

Thomas Jay Oord is Professor of Theology and Philosophy at Northwest Nazarene University. His most recent books include *Defining Love: A philosophical, scientific, and theological engagement* (2010) and *Creation Made Free: Open theology engaging science* (2005).

THE POLKINGHORNE READER

Science, faith and the search for meaning

Edited by
THOMAS JAY OORD

SPCK

TEMPLETON PRESS

Published in Great Britain in 2010 by
Society for Promoting Christian Knowledge
36 Causton Street
London SW1P 4ST
www.spckpublishing.co.uk

and in the United States of America in 2010 by
Templeton Press
300 Conshohocken State Road, Suite 550
West Conshohocken, PA 19428
www.templetonpress.org

British Library Cataloguing-in-Publication Data
A catalogue record for this book is available from the British Library

SPCK ISBN 978–0–281–06053–5

Library of Congress Cataloging-in-Publication Data is available
from the Library of Congress, Washington DC.

Templeton Press ISBN 978–1–59947–315–4

1 3 5 7 9 10 8 6 4 2

Typeset by Graphicraft Ltd, Hong Kong
Printed in Great Britain by Ashford Colour Press

Produced on paper from sustainable forests

For John and those who benefit from his ministry

IN CELEBRATION

*Celebrating eighty years of John Polkinghorne's life and his
contributions to science and theology*

Contents

Preface   ix

Introduction   1

Part 1
THE WORLD

1 The nature of science   9
2 The nature of the physical world   25
3 Human nature   36
4 The nature of reality   51
5 A brief history of science and religion   56
6 Science and religion as cousins   61
7 The work of love   68

Part 2
GOD

 8 The nature of theology   79
 9 Deity   88
10 Natural theology   94
11 Creation   104
12 Providence   110
13 Prayer and miracle   126
14 Time   133
15 Evil   137

Part 3
CHRISTIANITY

16 Scripture   147
17 The historical Jesus   159
18 The resurrection   176
19 Trinitarian theology   189
20 Eucharist   197

Contents

21 Eschatology   209
22 World faiths   225

Books by J. C. Polkinghorne   233
Bibliography   235
Index   241

Preface

I am very grateful to Thomas Jay Oord for this carefully integrated and wide-ranging selection from my writings. It conveys an accurate impression of my thinking on many central issues relevant to the interaction between science and religion. I particularly admire the way in which Dr Oord has succeeded in constructing a smoothly flowing text, avoiding the staccato character often to be found in a reader.

I hope that this volume will be of help to many who wish to engage with one of the most significant intellectual issues of our time: how one may take the insights of both science and religion with the seriousness that they demand, thereby gaining a deeper understanding of reality than either discipline could offer on its own.

John Polkinghorne
Queens' College
Cambridge

Introduction

John Polkinghorne

A vegetarian butcher.

That's the combination some people think analogous to someone being both a scientist and theologian. Yet John Polkinghorne is just that: a physicist and a priest. Combining the two in one person is likely to arouse curiosity and perhaps suspicion.[1]

Polkinghorne believes that together, both science and theology provide a particularly good vision of the world. Like the two scopes of a pair of binoculars that work in tandem to improve vision, science and religion offer essential perspectives for the great quest for truth.

Polkinghorne was born on 16 October 1930, in Weston-super-Mare in England. As a young boy, his high aptitude for mathematics and other subjects became evident. As an 18-year-old, he spent time in the military.

After military service, Polkinghorne entered Trinity College, Cambridge University, which had awarded him a scholarship. As an undergraduate there, he felt the call of God toward a deeper commitment to faith in Christ. His affirmative response to that call represented a significant moment in his Christian life. It was also while a student that he met the woman who would later become his wife, Ruth Martin, a student of Girton College.

Polkinghorne continued his studies at Cambridge, earning a Ph.D. in physics. He subsequently completed post-doctoral work in physics at California Institute of Technology in Pasadena, California.

His first teaching assignment was at the University of Edinburgh. After a few years at Edinburgh, however, Polkinghorne returned to Cambridge and served as professor in theoretical elementary particle physics. His contribution to research had primarily to do with the mathematical side of his discipline.

It was while a professor at Cambridge University that Polkinghorne wrote his first books. His first book about particle physics was called *The Particle Play*. The book 'gives an account of where we had found ourselves

---

[1] This idea and much of the following biographical information comes from John Polkinghorne's autobiography, *From Physicist to Priest: An Autobiography* (SPCK, 2007).

1

when the dust had finally settled on that phase of particle physics during which I had been an active participant,' Polkinghorne reports. A second book, *Quantum World* (1984), has proved the best-selling of the more than 35 volumes he has penned.

After almost two decades of research and teaching in physics, Polkinghorne decided to move in a very different direction with his life. In 1977, the year of his 47th birthday, he proceeded toward preparation for ordained Christian ministry in the Church of England. 'I simply felt that I had done my little bit for particle theory,' he explains, 'and the time had come to do something else.'[2]

Preparation for ministry, with the necessary study in theology, biblical studies and ministerial practice, brought Polkinghorne into greater appreciation for the essentials and nuances of Christian theology. The theologian whose writings were most influential on his journey was Jürgen Moltmann. 'The insight of the crucified God is at the heart of my own Christian belief,' he says in reference to Moltmann's famous book, *The Crucified God*, 'and indeed of the possibility of that belief.'[3]

The transition from working physicist to burgeoning priest led Polkinghorne to pen his newly forming ideas about the relationship between his Christian faith and science. The first of many books on the subject he titled *The Way the World Is*. The book addresses what would be a constant theme for Polkinghorne: the quest for the truth. He argues that both science and religion seek truth, but these kinds of truth are very different. 'In the one case it is truth about the impersonal, physical world open to our scientific manipulation and experimental interrogation; in the other case it is truth about the transpersonal reality of God, the One who is only fittingly to be encountered with awe and worship and obedience.'[4]

After several years working in the parish, Cambridge invited Polkinghorne back to serve as the Dean and Chaplain at Trinity Hall. However, after a short period at Trinity Hall, Queens' College, Cambridge, elected him as its president. He finished his career in this capacity and retired in 1996. Retirement from university life has allowed him to write and speak more frequently in public.

Polkinghorne has been a prolific interdisciplinary author. This kind of writing 'requires a degree of intellectual boldness and a degree of intellectual

[2] *Physicist to Priest*, 71.
[3] *Physicist to Priest*, 82.
[4] *Physicist to Priest*, 93.

charity', he says. 'I strive to be two-eyed, looking with both the eye of science and with the eye of religion, and such binocular vision enables me to see more than would be possible with either eye on its own.'[5]

This double focus also leads to a double mission. 'On the one hand,' says Polkinghorne, he tries in his writing 'to encourage scientists to take religion seriously and not dismiss it unreflectively without a hearing, and on the other hand to encourage religious people to take science seriously and not to fear the truth that it brings.'[6]

Polkinghorne is a critical realist. He believes that what we perceive can be a reliable guide to what is really the case. Who God is and the way God acts is in some way revealed to us. The way the Creator acts is greatly but not entirely different from how creatures act.

'If I can act in this way in a world of becoming that is open to its future,' argues Polkinghorne, 'I see no reason to suppose that God, that world's Creator, cannot also act providentially in some analogous way within the course of its history.'[7] This suggests that God provides some freedom and agency to creatures. 'God interacts with creatures,' he explains, 'but does not over-rule the gift of due independence which they have been given.'[8]

The theoretical aspect of the problem of evil has garnered significant attention from Polkinghorne. 'I believe that God wills directly neither the act of a murderer nor the devastation wrought by an earthquake, but both are permitted to happen in a world that is more than a divine puppet theatre.'[9] This not only means that creatures have free agency. It also means that the future is not already entirely settled. 'Even God does not yet know the unformed future, for it is not yet there to be known,' argues Polkinghorne, 'though undoubtedly God sees more clearly than any creature can the general way in which history is moving.'[10]

After an illustrious career of teaching, lecturing and writing, John Polkinghorne has been awarded a variety of honours. In 1993 and 1994, he was invited to give the Gifford Lectures at Edinburgh University. He chose to concentrate most of his lectures defending the cogency and fruitfulness of the Christian Nicene Creed. These lectures also provided a means by which Polkinghorne could elaborate the fecundity of his method of enquiry, what he calls 'bottom-up thinking'. By this, he means that he tries

---

[5] *Physicist to Priest*, 134.
[6] *Physicist to Priest*, 134.
[7] *Physicist to Priest*, 140.
[8] *Physicist to Priest*, 140–1.
[9] *Physicist to Priest*, 141.
[10] *Physicist to Priest*, 141.

not to prejudge what form an answer to a question ought to take. Instead, he seeks to look at the evidence and the kind of answer the evidence suggests. The lectures also provided opportunity to show scientists 'that they could engage with Christianity without the fear of being exhorted to commit intellectual suicide'.[11]

Polkinghorne is a Fellow of the Royal Society (the UK National Academy of Science). Not long after he retired in 1996, he was awarded the honour of Knight Commander of the Order of the British Empire. As an English priest, however, he follows the protocol of not using the usual prefix 'Sir' of a knight.

In 2002, Polkinghorne received the Templeton Prize for progress towards research and discovery about spiritual realities. This prize is monetarily the most valuable prize given each year. Polkinghorne also took the lead in founding the International Society for Science and Religion. The society is designed to encourage the deepest level of research in both theology and science.

The book's structure, style and notation

John Polkinghorne has done more than most – in fact, more than almost anyone – to address central issues in science and Christian theology. Most of his books, however, have been relatively short.

In the summer of 2007, I approached John with the idea of bringing together some of his best material from various small books. He agreed that the project needed to be done. So I proposed that we work together to construct this Polkinghorne reader. I am grateful to Templeton Press and SPCK, who signed on as co-publishers.

While I was confident that a book drawing together some of the very best from John's other books would be attractive to those who ponder issues in science and theology, I was equally confident that having John select the excerpts would be vital. In editing this book, therefore, I have relied heavily upon John's own preferences for which excerpts should be included.

The result is a book that presents to readers some of what John himself considers the best of his wide-ranging contributions to the field. Of course, I could not include absolutely everything of pertinence. However, the editorial choices made, most of which John suggested, reveal something of what possesses enduring relevance from Polkinghorne's prolificacy.

---

[11] *Physicist to Priest*, 145.

After much reflection, I divided this book into three segments: The World, God, and Christianity. These three overarching themes represent well the broad foci of Polkinghorne's research and writing. The excerpts in each segment are presented in such a way that the reader feels the flow of an argument and not an abrupt jump from excerpt to excerpt. This provides a coherent structure to the diverse topics addressed by the various chapter titles.

Readers will discover slight variations in spelling, punctuation and grammar. These differences arise from the differences between the grammar in the UK and the USA. For the most part, I have allowed these differences to stand without forcing uniformity. The differences are slight enough not to distract.

I also provide two bibliographies at the conclusion of the book. One provides a list of John Polkinghorne's books, from many of which this reader has drawn its material. The second bibliography lists books cited in the reader's excerpts.

I must acknowledge several people for their helpfulness with this project. I thank Natalie Lyons Silver, Laura Barrett and Lynn Coletta of Templeton Press, and Rebecca Mulhearn, Karen Beerman and Philip Law from SPCK. I thank Karl Giberson and Dean Nelson for their help in various aspects of the project. I thank my assistants Jill Jones, Dannea Miller and Andrew Schwartz.

When I think about the research occurring at the science and Christian theology interface and John Polkinghorne's particular work in this regard, I am reminded of the words of the Apostle Paul: 'What can be known about God is plain to them, because God has shown it to them. Ever since the creation of the world his eternal power and divine nature, invisible though they are, have been understood and seen through the things he has made' (Rom. 1.19–20, NRSV).

I trust this reader serves as a witness both to the best of John's writing and to the truths the Creator reveals.

Thomas Jay Oord

Part 1
THE WORLD

1

The nature of science

There is a popular account of the scientific enterprise which presents its method as surefire and its achievement as the inexorable establishment of certain truth. Experimental testing verifies or falsifies the proposals offered by theory. Matters are thus settled to lasting satisfaction; laws which never shall be broken are displayed for all to see.

In actual fact, as we shall find out, the matter is a good deal subtler than that. Nevertheless, the great enhancement that the twentieth century has seen in our understanding of the world in which we live, even encompassing an account of its earliest moments 14,000 million years ago and including the beginnings of a comprehension of how life could have evolved from inanimate matter, together with the remarkable technological developments stemming from scientific advance, lends a certain credibility to this triumphalist point of view. Such splendid successes suggest that here is the key to real knowledge. In the bright light of science's achievements, other forms of discourse are in danger of appearing mere expressions of opinion. The widespread thought that science has somehow 'disproved religion' is based on psychological effect rather than logical analysis. It is a continuation of the Enlightenment distrust of all knowledge which is not patterned according to the paradigm of scientific method.

It is ironic that at the same time that there is this widespread popular attitude there is also, in circles more austerely intellectual, a critical review of the nature of the scientific method and of its actual achievement. The practices of science have been reassessed and its procedures found to be more complex and questionable than the simple popular account acknowledges. The picture of the professor in his laboratory watching the pointer move across the scale to the expected reading, and thereby establishing his theory beyond the possibility of doubt, bears about as much relation to reality as does the simplicity of the comic-strip detective to the complexities of actual police investigation. If the method of science is open to revaluation, so, of course, will be the nature of the conclusions resulting from it. It is to these matters that we must now turn.[1]

* * *

In order scientifically to interrogate the world, we have to do so from a point of view. It is precisely this need for an (admittedly corrigible) theoretical expectation which distinguishes science from its precursor, natural history, which is simply content to take in the flux of apparent experience as it happens. In a famous phrase, Russell Hanson referred to this theory-laden character of our observation as 'the spectacles behind the eyes.'[2] Our scientific seeing is always 'seeing as.'

To recognize this is to raise the question of the character of our experimental knowledge. The role of observation as the stern and impartial arbiter of scientific theory is somewhat compromised if in fact the image of nature we receive is always refracted by those spectacles behind the eyes. Might there not be a variety of possible perspectives on the world of which the received scientific view at any time is just one option?

In books on the philosophy of science, this possible dilemma is often illustrated by the notorious duck/rabbit, a sketch which, looked at one way, can be seen as a duck and which, looked at another way, can be seen as a rabbit, the open bill of one becoming the ears of the other. Actually, this particular ambiguity is rather readily resolved by acknowledging that what is before us is a rather exiguous line drawing. Physics itself provides a much more striking example of such ambivalence.

The conventional view of quantum theory,[3] accepted by the vast majority of physicists, states, for example, that there is no assignable cause for the decay of a radioactively unstable nucleus at any particular moment. All that can be asserted is that there is a calculable probability for such a decay taking place within a specified period of time. The quantum physicist is in the same practical position as the actuary of a large insurance company who is unable to say whether any particular client will die in the coming year, but who can be tolerably sure that a calculable number of clients in a particular age group will die within that period. However, there is an important difference between the physicist and the actuary,

---

[1] The preceding section is from *One World* by John Polkinghorne © 1986 by John Polkinghorne, revised edition published by Templeton Press, 2007, 9–10.

[2] N. R. Hanson, *Perception and Discovery* (San Francisco: Freeman Cooper, 1969), ch. 9.

[3] For details of these matters, see J. C. Polkinghorne, *The Quantum World* (Longman, 1984), ch. 5.

according to conventional quantum theory. There are causes why the actuary's clients die, even if they are not known to him. There are asserted to be no causes for individual events in the quantum world.

To this conventional quantum interpretation, there is an alternative point of view, first worked out successfully by David Bohm. It asserts that all events are causally determined, but some of these causes (called in the trade 'hidden variables') are inaccessible to us. That is the reason, in Bohm's view, why our actual knowledge has to be statistical. It is a matter, not of principle, but of ignorance. This point of view is, of course, identical with that of the actuary, whose clients die of causes, to him unknown.

In the realm of non-relativistic quantum theory (that is, concerning the behavior of very small and slowly moving systems), the conventional theory and Bohm's theory give exactly the same experimental results. Yet the understandings they offer are radically different. Here is a duck/rabbit with a vengeance! Why then do the majority of physicists believe the one in preference to the other? It is clearly not a matter of observational decision.

I think there are two reasons for the majority preference for conventional quantum theory (which I share). The first is that Bohm's theory, though very clever and instructive, has a contrived air about it. It is significant that this is enough to put off most professionals despite the theory's 'common sense' determinism, which might seem an overwhelmingly attractive feature to a layman. Matters of taste, judgments of elegance and economy, play an important part in the development of science. By these canons conventional quantum theory seems to most of us more elegant, and so more compelling, than Bohm's ingenious ideas.

But why should the more elegant prove scientifically the more compelling, other things being experimentally equal? Here we see the coming into play of a factor, the search for simplicity, which goes beyond the impersonality of the popular account of the scientific enterprise. After all, is not one man's simplicity another man's complication? Does it not all depend on those spectacles behind the eyes?

To Copernicus as much as to Ptolemy, the circle was the perfection of simplicity. It was only natural, in their view, that heavenly motion should be explained in circular terms. Kepler's introduction of ellipses must have seemed to many of his contemporaries a most ugly and unwelcome development. Simplicity only returned to celestial mechanics with the totally different beauty of the inverse square law inserted into Newtonian dynamics.

Today we retain a belief in the elegance and economy of gravitational physics, though its current expression would be in terms of the geometrical curvature of space-time described by Einstein's general relativity (if one uses the language of classical physics) or in the gauge theory of massless gravitons (if one uses the language of quantum theory). Beauty is indeed in (or behind) the eye of the beholder. Its influence on scientific thought is undeniable, but that very statement raises the question of the true nature of that thought.[4]

* * *

The simple account of science sees its activity as the operation of a methodological threshing machine in which the flail of experiment separates the grain of truth from the chaff of error. You turn the theoretic-experimental handle and out comes certain knowledge. The consideration of actual scientific practice reveals a more subtle activity in which the judgments of the participants are critically involved.

If you wish to give an experimental physicist an uneasy moment, look him straight in the eyes and say, 'Are you sure you have got the background right in your latest experiment?' (In other words, 'Are you sure you have eliminated all possible sources of spurious effects and are actually measuring what you claim to measure?') If you wish to give a theoretical physicist an uneasy moment, look him straight in the eyes and say, 'That latest theory of yours looks a little contrived to me.' (In other words, 'I do not see in it that look of elegant inevitability which time and again has proved the hallmark of true theoretical insight.') Their answers will not depend upon simple ineluctable prediction confronting indisputable fact. Rather, they will involve a reasoned discussion of how those concerned evaluate and interpret the situation.

This role of personal judgment in scientific work was emphasized by Michael Polanyi.[5] He called it tacit skill. Acts of discrimination are called for in concocting a successful scientific theory which are no more exhaustively specifiable than are the skills of a wine-taster in blending a good sherry. But just as the sherry blender has to submit the result of his labors to the judgment of the discerning public, so the scientist has to persuade his colleagues of the soundness of his judgment. This necessity saves personal knowledge from degenerating into mere idiosyncrasy.

---

[4] The preceding section is from *One World* by John Polkinghorne © 1986 by John Polkinghorne, revised edition published by Templeton Press, 2007, 12–15.

[5] M. Polanyi, *Personal Knowledge* (London: Routledge & Kegan Paul, 1958).

Once one has acknowledged the part that personal discrimination has to play in scientific endeavor, the whole enterprise may seem to have become dangerously creaky, its rationality diminished or even destroyed, by the importation of acts of individual judgment, even if they are claimed to be validated by the eventual assent of the scientific community. Has not the austere search for truth degenerated into the proclamation of an ideology, even if democratically endorsed by its adherents? There have certainly been philosophers of science who have taken such a view, and it is from them that the scientific method has received its most severe criticism.

Thomas Kuhn studied those rare moments in the history of science when a major change occurs in the scientific worldview. Most of the time, scientists are engaged in problem-solving, applying an agreed over-all understanding to the attempt to explain particular phenomena. Just occasionally, however, it is the overall understanding itself which is subject to radical revision.

An example of such a paradigm shift, as Kuhn calls it, would be the transition from classical to relativistic dynamics. For Newton there is a universal uniformly flowing time; for Einstein each observer experiences his own time so that two observers in relative motion will not agree about which events are simultaneous with each other. For Newton a particle's mass is an unchanging quantity; for Einstein it varies with the motion of the particle.

Clearly there is a striking difference between these two systems of mechanics. We can all agree on that. But Kuhn proclaims a divorce between the two so absolute that he can say, 'In a sense that I am unable to explicate further, the proponents of two competing paradigms practice their trades in different worlds.'[6] This is his celebrated claim that two competing paradigms, such as Newtonian and Einsteinian mechanics, are incommensurable; that is, there is no point of contact and comparison between them. If this were really so, it would imply that there were also no rational grounds for preferring one to the other, since such grounds would depend on the possibility of making just such a critical comparison. Kuhn does not flinch from drawing that conclusion:

> As in a political revolution, so in paradigm choice – there is no standard higher than the consent of the relevant community. To discover how scientific revolutions are effected, we shall therefore have to examine not only the

---

[6] T. Kuhn, *The Structure of Scientific Revolutions*, 2nd edn (Chicago: University of Chicago Press, 1970).

impact of nature and of logic, but also the techniques of persuasive argumentation effective within the quite special groups that constitute the community of scientists.[7]

Thus Kuhn's study of scientific revolutions has led him to accentuate the role of the personal factor to the extraordinary extent of proclaiming the efficacy of scientific mob rule.

All this is really very curious and greatly overdone. Did special relativity really come to be adopted because Einstein had a propaganda machine superior to that of Lorentz? Experimental evidence (such as the eventual confirmation of the slowing of moving clocks via observation of the lifetimes of rapidly moving particles) presents perfectly adequate nonideological reasons for accepting the theory. While Newton's and Einstein's understandings of mass are very different, is there not sufficient residual common ground for us to be able to say that they are offering alternative, and so comparable, accounts of inertia? Kuhn dismisses as an irrelevancy the well-known fact that Newtonian mechanics is the slow-moving limit of Einstein's mechanics. Yet to physicists this relationship would seem to be important, for it explains why classical mechanics was so long an adequate theory and why it remains so for systems whose velocities are small compared with the velocity of light.

Of course, study of persuasive techniques can help us understand why a new scientific viewpoint gains quick or slow acceptance. However, to suppose that this provides the major part of the story of how new ideas are embraced is surely preposterous. Indeed, in later writings Kuhn himself seems to have withdrawn from so extreme a position.

Kuhn's revolutionary incommensurability, if true, would undermine the idea that science can claim our rational, as opposed to rhetorical, assent. An even stronger threat to that idea is posed by the writings of Paul Feyerabend. He is a philosophical enfant terrible who does not hesitate to proclaim that the scientific emperor has no methodological clothes. Our discussion of skill has made it clear that there is no totally specifiable set of rules for scientific theory choice. There is no algorithmic machine, the turning of whose handle is guaranteed to lead to the Nobel Prize. At best, there are only guiding principles, exercised with discrimination by experts whose conclusions are subject to the collective judgment of the scientific community. Feyerabend seizes on this tacit, unspecifiable element and blows it up into a dominating principle of scientific laissez-faire. He claims that in science 'the only principle that does not inhibit progress is *anything*

---

[7] Ibid., 94.

goes.'[8] He is a self-proclaimed scientific anarchist. What in Kuhn was simply preposterous becomes in Feyerabend the Theatre of the Absurd.

If science is an intellectual free-for-all, then there is no reason for preferring astronomy to astrology, the oxygen theory of combustion to the phlogiston theory. Feyerabend honestly recalls that 'having listened to one of my anarchic sermons, Professor Wigner [a distinguished theoretical physicist] replied "But surely you do not read all the manuscripts that people send you, but you throw most of them into the wastepaper basket."' He acknowledges that he does so, but 'partly because I can't be bothered to read what does not interest me . . . partly because I am convinced that Mankind, and even science, will profit from everyone doing his own thing.'[9] His impishness is irrepressible.

Yet another assault on the rationality of science is mounted by adherents of what is called the 'strong program' in the history of science. They assign to social forces a prime causative role in scientific change. For example, Andrew Pickering wrote of the recent sequence of investigations in high energy physics which have led physicists to believe that matter is composed of quarks and gluons, 'The world of HEP [High Energy Physics] was *socially* produced.'[10] The claim is that the largely unconscious adoption of certain conventions of experimental interpretation, together with a collective expectation framed in particular theoretical terms, has so molded the thought of the invisible college of high energy physicists that a quark model of matter was imposed on the supposedly plastic mass of available data. The assertion is there in the title of his book; it is 'constructing' quarks, not 'discovering' them.

Weighty grounds would be required for so startling a conclusion. On investigation we find that all that is offered is an analysis of such incidents as the differing background calculations, which in the 1960s seemed to exclude neutral currents, but which in their revised form were the basis for confirming the by then theoretically acceptable neutral current in the 1970s. We can readily agree that this is an excellent example of how social forces can retard or accelerate the pace of scientific discovery, but there are no grounds at all for going on to assert that they actually control the nature of that discovery. After all, the dust does settle. No one would now claim that the neutral current is an artifact of background calculations. The cumulative weight of evidence for its existence, made clearer by

---

[8] P. Feyerabend, *Against Method* (London: Verso, 1975), 23.
[9] Ibid., 215.
[10] A. Pickering, *Constructing Quarks* (Edinburgh: Edinburgh University Press, 1984), 406.

increased understanding of how to perform the calculations of neutron-induced background, has simply settled that issue.[11]

* * *

The recognition of a role for judgment in the scientific enterprise, a tacit element not wholly reducible to the application of rules specifiable a priori, gives it a kindred character to aesthetic, ethical, and religious thinking. Many have asserted these latter modes of thought to be of a different and inferior kind, matters of mere opinion. We now see that what is involved in the comparison is a question of degree rather than an absolute distinction. To say so is not, as Kuhn and Feyerabend and others have suggested, to open the door to irrationality. It is simply to recognize that reason has a broader base than corresponds to a totally specifiable method of verification. (The qualification *total* is vital here; we are not saying that anything goes.) The mind has its reasons that computers know not of. The justification of this rational claim depends, I believe, on an assessment of the actual nature of the scientific achievement. By their fruits ye shall know them. We must consider what those fruits actually are.

At first sight the prospect might seem discouraging. Paradoxically, the advancing success of science appears subversive of its attainment of truth. Do not all theories in the end prove inadequate and have to be replaced? We once thought that the basic constituents of matter were atoms; then nuclei; then protons and neutrons; then quarks and gluons; next – maybe strings? Bigger fleas have lesser fleas, and so ad infinitum. Isn't ultimate scientific truth a will-o'-the-wisp? Newton-Smith calls this the pessimistic induction: 'any theory will be discovered to be false within, say, two hundred years of being propounded.'[12] There is excellent evidence for adopting this maxim. However, to do so is only fatal if we thought that certain truth is our necessary goal. In fact we shall have to be content with the more modest aim of verisimilitude. Our understanding of the physical world will never be total, but it can become progressively more accurate.

The analogy of a sequence of maps of increasingly larger scale may be helpful. None will ever tell us all there is to be told about that particular piece of terrain. Each is a kind of coarse-grained isomorphism, representing

---

[11] The preceding section is from *One World* by John Polkinghorne © 1986 by John Polkinghorne, revised edition published by Templeton Press, 2007, 16–20.

[12] W. H. Newton-Smith, *The Rationality of Science* (London: Routledge and Kegan Paul, 1981), 14.

accurately features from a certain size upwards but ignoring or smoothing out those which are smaller. For different purposes different maps are adequate. As a motorist I do not need the detail I would demand as a hiker. In the same way established scientific theories do not disappear; they simply have their domain of applicability circumscribed. Newtonian mechanics is satisfactory for largish objects moving at ten miles an hour, unsatisfactory for the same objects moving at a hundred thousand miles a second. Scientific theories are corrigible, but the result is a tightening grasp of a never completely comprehended reality.

So I would wish to say. But there are many who would deny it. First, there is the problem of how we know, even within a prescribed domain, that we have arrived at an adequate map of its physical behavior. Newtonian mechanics has so far proved excellent for describing the collisions of billiard balls, but how can I be sure that the next time I approach the table they will not be found to be behaving differently? Anyone attempting to make a general statement faces the problem of induction, of how to produce universal laws from the study of specific instances. One cannot examine every electron in the universe before saying anything about electrons in general. Nor can one survey every billiard ball collision that ever has been, or ever will be, before pronouncing on how such objects behave. That resolute skeptic, David Hume, was the first to emphasize the logical difficulty this presents. How then can science proceed?

One possible response is to moderate the claim. This is the attitude of Karl Popper. He exhibits a maximal distrust of induction. However many 'for instances' there may be in favor of a theory, there is always an infinity of untried cases in which it might prove wrong. The odds are thus permanently stacked against its validity. In Popper's view, therefore, we abandon all hope of verification. The best that can be done is to settle for falsifiability. While any number of successes will never count in a theory's favor, one failure will prove fatal. 'Only the falsity of the theory can be inferred from empirical evidence, and this inference is a purely deductive one.'[13] This chilling message is conveyed in the original with the emphasis of italics. Clearly if that is all that can be said, the nature of the scientific enterprise is precarious indeed. 'The empirical basis of objective science has then nothing "absolute" about it. Science does not rest upon rock-bottom,' says Popper. 'The bold structure of its theories rises, as it were, above a swamp.'[14]

---

[13] K. Popper, *Conjectures and Refutations* (New York: Basic Books, 1962), 55.
[14] K. Popper, *The Logic of Scientific Discovery* (London: Hutchinson, 1968), 111.

Popper is driven to this gloomy assessment because he exalts logic above intuition. (It is only the 'deductive' which is safe.) Since the ratio of the number of successful answers to the number of potential questions is inevitably, for any theory, a finite number over infinity, those who rely on such wary calculation of odds will always condemn themselves to a state of intellectual pessimism. It is imposed by their timidity. Newton-Smith says, 'One cannot over-stress the counter intuitive character of [Popper's] position.'[15] Is the Newtonian mechanics of billiard balls really in a state of permanent jeopardy? I think not.

Those who balk at induction do so because there is no exhaustively specifiable set of rules which enable one to lay down a priori when its application is justifiable. Its employment involves an act of judgment, even though in the case of a theory well tried in a definite domain, such as Newtonian mechanics, one cannot feel that great powers of discrimination are required for its successful exercise. We have already recognized that such acts of judgment enter other aspects of the scientific enterprise. That being so, an answer to the Humean criticism which is preferable to the partial surrender of Popper is simply to assert that we shall rely upon inductive method exercised with an appropriate degree of skill. Undoubtedly that attitude corresponds to the actual practice of science, and it seems to have stood the subject in good stead.

Science certainly appears successful. It has the air of progress about it. But what exactly is the nature of its achievement? Here we come to the second set of objections to any claim that science results in a tightening grasp of a never completely comprehended reality. It is asserted that the use of that last word is a naive misapprehension of what science is actually about. We have reached the parting of the ways between the positivists, the idealists, and the realists.

Those of a positivist persuasion lay stress on perceptions which can be intersubjectively agreed; the scientific task is the harmonization of such experience. Entities not directly accessible to experience, such as electromagnetic fields and quarks, which form the staple of the discourse of fundamental physics, are said to be just manners of speaking which are useful simply as means to that reconciling end. They do not represent actually existing realities. The scientific world is populated by pointers moving across scales and marks on photographic plates, rather than potentials or electrons; theories are just convenient summaries of data.

---

[15] Newton-Smith, *Rationality of Science*, 62.

There are, in fact, very considerable difficulties in drawing that clear distinction between the facts of data and the devices of theory which my simple summary of positivism has assumed. The objections to positivism, however, go beyond that. Its arid account seems totally inadequate to explain the actual practice of science. After all, if all that happens is the reconciliation of various bits of experience, much of it recondite, why is it worth all the painful labor involved? Bernard d'Espagnat, speaking of the activity of elementary particle physicists, wrote:

> Whereas the activity appears essential as long as we believe in the independent existence of fundamental laws which we can still hope to know better, it loses practically its whole motivation as soon as we believe that the sole object of the scientists is to make their impressions mutually consistent. These impressions are not of a kind that occur in our daily life. They are extremely special, are produced at great cost, and it is doubtful that the mere pleasure their harmony gives to a selected happy few is worth such large public expenditure.[16]

Or, I would add, the dedication and toil of those involved.

The philosophical problems of positivism, together with its impoverished account of scientific motivation, mean that it has few adherents in its pure form. However, there are accounts of the nature of scientific achievement, less inadequate than positivism but substantially influenced by it, which make claims that fall short of the realism I am wishing to defend.

Science is concerned with the power to predict or the power to manipulate phenomena, we are told. These two abilities are closely connected, for to foresee is to be forewarned and so at least to some extent to be in a position to take action to obtain a desired outcome. Science does indeed manifest such instrumental capacity, but should we be content with that and not go on to claim that its final goal is understanding?

An instrumentalist would maintain that the only question to ask about a theory is, 'Does it work?' If it does, we are not to bother whether it is true or not. The suggestion urged on Galileo by Cardinal Bellarmine, that the Copernican system was just a means of 'saving the appearances' (of getting the answers right) but did not describe how things actually were, would be endorsed enthusiastically by someone of this persuasion. However, it will not do.

Suppose that the Meteorological Office was given a sealed machine which had the property that if you fed in details of today's weather, the

---

[16] B. d'Espagnat, *The Conceptual Foundations of Quantum Mechanics* (Menlo Park, Calif.: Benjamin, 1971), 474.

machine would correctly predict the weather for any day ahead in the following year. The predictive role of the Met Office in weather forecasting would be perfectly fulfilled. Would that mean that all its meteorologists would simply pack up and go home? Not at all! They are also interested in understanding the way in which the earth's atmosphere and the sea and the landmasses interact as a giant heat engine to produce our climate. Before long some of them would be tampering with the seals on the machine in the hope of finding out how it worked, expecting that that would lead to an improved comprehension of the weather system that it modeled so accurately.

No account of science is adequate which does not take seriously this search for understanding, together with the experience of discovery which vividly conveys to the participants the impression that understanding is what they are actually attaining. I have never known anyone working in fundamental physics who was not motivated by the desire to comprehend better the way the world is. It is because they yield understanding, though often having low or zero predictive power, that theories of origins, such as cosmology or evolution, are rightly classed as parts of science.

To claim that understanding is the true goal of science and the nature of its actual achievement is not of itself to have reached the realist position I wish to defend. We have to ask the further question of where this understanding comes from. Is it imposed by us, or is it dictated by the nature of the world with which we interact?

The former account would be given by those who take an idealist position. The modern grandfather of this point of view is Immanuel Kant, who believed that space and time are necessary mental categories which we impose on the flux of experience in order to be able to cope with it at all. This kind of view has not been without its supporters in the scientific community.

Sir Arthur Eddington, in a famous parable, compared physicists to fishermen using nets with a certain width of mesh, who concluded that there were no fish in the sea smaller than that particular size. In other words, the apparent ordered reality that we think we perceive is alleged to be the product of our observational procedures.

The American physicist Henry Margenau was bold enough to admit the consequences of such ideas and said, 'I am perfectly willing to admit that reality does change as discovery proceeds.'[17] In his view the neutron did not exist prior to its 'discovery' in 1932. One's feeling that such a statement

---

[17] H. Margenau, *The Nature of Physical Reality* (New York: McGraw-Hill, 1950), 295.

is, to say the least, highly unsatisfactory is reinforced by a consideration of the track record of idealist claims.

Kant believed that he had demonstrated that space had to be three-dimensional Euclidean in structure. With our knowledge of non-Euclidean curved spaces, actually realized in general relativity, we can see that all that he succeeded in doing was to produce a specious rationalization of what at the time was thought to be the only physical possibility.

Eddington spent the last years of his life developing the tortuous ideas published posthumously in his book *Fundamental Theory*.[18] Its supposedly rationally established conclusions have signally failed to correspond to the structure of the physical world revealed to subsequent investigation. If fruitfulness for the future is a good test of scientific creativity, idealist notions have proved a dismal failure.

We need not be surprised. The world, though ordered, is strange and subtle. Our powers of rational prevision are pretty myopic and limited by the contingency of the way things are, existing independently of how we think they ought to be. The natural convincing explanation of the success of science is that it is gaining a tightening grasp of an actual reality. The true goal of scientific endeavor is understanding the structure of the physical world, an understanding which is never complete but ever capable of further improvement. The terms of that understanding are dictated by the way things are.

That is the realist position. It certainly corresponds to the way scientists themselves see their activity and are encouraged to persevere with it. Of course, most of them are philosophically unreflective people, and it might be that this is just a shared naive misapprehension. Yet the way devotees of a subject view their practice must surely count for something in its evaluation. Many philosophers of science have been unwilling to give this due recognition, feeling that they knew best, without paying sufficient attention to what the honest toilers had to say. The realist view, it seems to me, is the only one adequate to scientific experience, carefully considered.

If realism is to prove defensible it has to be a critical, rather than a naive, realism. First, it has to recognize that at any particular moment verisimilitude is all that can be claimed as science's achievement – an adequate account of a circumscribed physical regime, a map good enough for some, but not for all, purposes. Once one moves outside regimes already explored, to hitherto unattained high energies for example, there is every prospect

---

[18] A. S. Eddington, *Fundamental Theory* (Cambridge: Cambridge University Press, 1946).

that modification of our theories will be required to take account of unforeseeable phenomena. These modifications may, at times, be drastic (as when Einstein takes over from Newton), but there is sufficient residual continuity to discount the Kuhnian claim that we have lurched from one world to another, disjoint from it.

Second, our everyday notions of objectivity may prove insufficient as we move into regimes ever more remote from familiar experience. Quantum theory presents us with exactly this happening. According to Heisenberg's uncertainty principle, for entities like electrons we cannot know both where they are and what they are doing. This abolishes picturability in the quantum world. Realism is not tied to simple notions derived from everyday experience alone.

Third, a critical realism is not blind to the role of judgment in the pursuit of science. It acknowledges that the simple picture of definite theoretical prediction confronting unquestionable experimental fact and leading to certain truth is too unsubtle an account of what science is about. As Newton-Smith says, 'The story of SM [Scientific Method] will not produce a methodologist's stone capable of turning the dross of the laboratory into the gold of theoretical truth.'[19] There are always unspecifiable discretionary elements involved.

We cannot take off our spectacles behind the eyes. However, if experiments are theory-laden, it is also true, as Carnes points out, that theories are fact-laden.[20] They are responses to what is perceived to be there and in need of explanation. Perhaps the most troublesome question for the critical realist arises from the fact that for any finite set of data, there will always be a variety of possible theories which could fit it. (One could call this the duck/rabbit problem.) A rational criterion of choice is provided by demanding that an acceptable theory should prove its fruitfulness. It can do so in two ways: by a capacity to continue to cope with data as their range and accuracy expands, and by the theory being shown to have correct conclusions unforeseen at the time of its devising.

As an example of the former, consider the Newtonian account of the solar system. For about two centuries every new result coming from increased observational accuracy could be explained by a natural refinement of calculational technique. These theoretical responses represented fine-tuning in accuracy (for example, by taking into account hitherto neglected interplanetary effects) which was wholly in accord with the

---

[19] Newton-Smith, *Rationality of Science*, 209.
[20] J. R. Carnes, *Axiomatics and Dogmatics* (New York: Oxford University Press, 1982), 14.

spirit of the theory and in no sense imposed upon it. (In contrast, a stubborn adherent of the Ptolemaic theory would have had to introduce ad hoc a new set of epicycles every time better observations were available.) The most striking illustration of such natural development of Newtonian ideas was provided by the work of Adams and Leverrier. They explained perturbations in the orbit of Uranus by supposing them to be due to a further, and till then unknown, planet. Their suggestions were triumphantly confirmed by the discovery of Neptune. The power of a theory to respond to progressive experimental probing without arbitrary manipulation is strong evidence of its verisimilitude. We cannot go on to say its truth, because its fruitfulness is not unlimited. An unresolved small discrepancy in the advance of the perihelion of Mercury eventually showed that even the Newtonian theory of gravitation had its limited domain of applicability. The explanation of this phenomenon required Einstein's general theory of relativity.

As an example of the second type of fruitfulness, we can consider Dirac's theory of the electron. In 1928 he devised an equation which successfully combined quantum mechanics with special relativity. Such a nontrivial synthesis was necessary to describe particles which are small and fast moving. It was an unexpected bonus when it was found that the same equation also explained the fact, till then mysterious, that the electron's magnetic properties were twice as strong as one would naively have expected. When this sort of thing happens, it is very convincing evidence for the verisimilitude of the theory. Again, it was no more than that, for it was eventually found that there are small corrections to the electron's magnetic behavior which require for their explanation the much more elaborate theory called quantum electrodynamics.

I believe that, after a certain time of testing, theories which gain wide acceptance in the scientific community have exhibited their reasonableness by demonstrating just such fruitfulness. Such rational staying power conveys an impression of naturalness and lack of contrivance which is convincing. Thus the underdetermination of theory by data does not pose a fatal difficulty for realism, since the theories which survive have been selected by the rational criterion of sustained success. Nor do I think that the lack of effective competing theories is to be attributed to a slothful acquiescence in a socially induced consensus. Scientists are active in a continual attempt to devise alternatives to received opinion, impelled not only by the search for truth but also by the desire to establish personal reputation.

I have attempted to defend a view of science which asserts its achievement to be a tightening grasp of an actual reality. In the course of the discussion, we have acknowledged the role that personal judgment, presented for the approval of the community and pursued along lines which are rational but not wholly specifiable, has to play in the enterprise. In my view this means that science is not different in kind from other kinds of human understanding involving evaluation by the knower, but only different in degree. It is clear that the personal element is less significant in science than in, say, judging the beauty of a painting, but it is not absent. We are to take what science tells us with great seriousness. However, we are not to assign it an absolute superiority over other forms of knowledge so that they are neglected, relegated to the status of mere opinion. Our discussion has taken science off the pedestal of rational invulnerability and placed it in the arena of human discourse. It is not the only subject with something worth saying. If differing disciplines, such as science and theology, both have insights to offer concerning a question (the nature of humanity, for example), then each is to be listened to with respect at its appropriate level of discourse.[21]

---

[21] The preceding section is from *One World* by John Polkinghorne © 1986 by John Polkinghorne, revised edition published by Templeton Press, 2007, 21–31.

2

The nature of the physical world

'There is no sense in which subatomic particles are to be graded as "more real" than, say, a bacterial cell or a human person, or even social facts.'[1] The words are those of that resolute antireductionist, Arthur Peacocke. In a series of writings, he has defended the existence of level autonomy in our descriptions of the physical world. Biology has its own concepts and understandings which are not reducible to complicated corollaries of physics and chemistry.[2] I certainly agree that this is so.[3] Yet it is hard indeed to dispel altogether from one's thinking a certain reductionist tendency.

When we start to consider the nature of physical reality, it is instinctive to turn first to the insights of so-called fundamental science, to start with elementary particle physics and its spatially big brother, cosmology. Our discussion then becomes one of 'emergence': how, within physics itself and beyond it, new properties arise – such as the power of 'classical measuring apparatus' to determine the outcome of uncertain quantum mechanical experiments; the ability of complex molecules to replicate themselves; the coming to be of consciousness, self-consciousness, worship.

In actual fact, we understand very little of how these different levels relate to each other. The problems are mostly too hard for current knowledge, despite the stunning successes of molecular biology in casting light on the physical basis of genetics. But the direction in which to look for an understanding seems clear enough. It will come from being able to relate the higher level to the lower. Emergence is conceived as a one-way process, by which the higher whole arises from the complex organization of its lower parts.

The reasons for thinking this way appear clear enough. Vitalism seems dead and even the most fervent antireductionist in relation to concepts accepts a structural reductionism. Physical reality is made out of the

[1] A. R. Peacocke, *God and the New Biology* (London: Dent, 1986), 28.
[2] A. R. Peacocke, *Creation and the World of Science* (Oxford: Oxford University Press, 1979), ch. 4; Peacocke, *God and the New Biology*, chs 1 and 2.
[3] J. C. Polkinghorne, *One World* (SPCK, 1987), ch. 6.

entities described by fundamental physics – quarks and gluons and electrons (or superstrings, or whatever). Hence, the feeling that if one day we wrote the equations of a Theory of Everything on our T-shirts, we should have got somewhere, despite the fact that in terms of our actual understanding of the physical world those equations would be more like the precise statement of the problem, rather than its solution.

Another encouragement to such a bottom-up way of thinking is that it recapitulates the way in which we believe the actual complexity of being to have come about. First there was the quark soup of the primeval universe; then nuclear matter after those famous first three minutes;[4] then simple atoms when the background radiation was 'frozen' out after about half a million years; much later the complex molecules in the shallow seas of early Earth; then unicellular life; then animals; then *homo sapiens*. 'Ontogeny recapitulates phylogeny,' not only embryologically but also conceptually.

Yet it is possible that if subatomic particles are not 'more real' than cells or persons, they are not more fundamental either. It is possible that emergence is, in fact, a two-way process; that it would be conceptually valid and valuable to attempt to traverse the ladder of complexity in both directions, not only relating the higher to the lower but also the lower to the higher. Such a proposal goes somewhat beyond the mere acknowledgement of level autonomy, for it suggests the existence of a degree of reciprocity between levels.

I am tempted to explore this notion because of a recent development in physics itself. I refer to that theory of complex dynamical systems which goes under the not altogether appropriate name of the theory of chaos.[5]

* * *

It is characteristic of chaotic systems generally that unless one knows the initial circumstances with *unlimited* accuracy, one can only project their behaviour a small way into the future with any confidence. Beyond that they are intrinsically unpredictable.

It will not surprise you to learn that this feature of chaotic unpredictability first came to light during computer investigations of weather forecasting, using simple models of the behaviour of the atmosphere. It

---

[4] S. Weinberg, *The First Three Minutes: A Modern View of the Origin of the Universe* (New York: Basic Books, 1993).

[5] The preceding section is from *Reason and Reality* by John Polkinghorne © 1991 by John Polkinghorne, published in the USA by Trinity Press International, 1991, 34–5.

gives rise to 'what is only half-jokingly known as the Butterfly Effect – the notion that a butterfly stirring the air today in Peking can transform storm systems next month in New York.'[6] Yet there is also a contained random-ness about the behaviour of chaotic systems. They do not wander all over the place but their motions home in on the continual and haphazard exploration of a limited range of possibilities (called a strange attractor). There is an orderly disorder in their behaviour. That is why chaos theory was not a well-chosen name.[7]

* * *

The resulting worldview is certainly not that of a dull mechanical regularity. Indeed, the behaviour envisaged has more than a touch of the organismic about it. This feeling is reinforced by consideration of other insights into physical process that have been gained in recent years. I am thinking of the study of dissipative systems, whose behaviour has been a major topic for investigation by Ilya Prigogine and his collaborators.[8] These systems are maintained far from equilibrium by an inflow of energy from the environment. The spontaneous triggering effects of small fluctuations, too tiny to be directly discernible, induce an order which is maintained by the flow of energy. The red spot of Jupiter, which has maintained its shape for centuries amidst the turbulent eddies of that planet's atmosphere, is thought to be an example. The order thus supported may be dynamic-ally changing, as in the case of the so-called chemical clock. With a care-fully controlled steady inflow and outflow of materials, the chemical constituents present in a mixture are found in certain circumstances to perform rhythmic oscillations from one concentration to another and back again, an astonishing effect involving the 'collaboration' of trillions of molecules. In this kind of phenomenon one sees the generation of novel and large-scale order which seems quite incomprehensible at the micro-scopic molecular level. Physics is found to describe processes endowed not just with being but also with becoming.

The physical systems about which I have been talking are complicated, but they fall far short of the complexity of even the simplest living cell. Its biochemical dance also exhibits the combination of openness and order

---

[6] J. Gleick, *Chaos* (London: Heinemann, 1988), 8.

[7] The preceding section is from *Reason and Reality* by John Polkinghorne © 1991 by John Polkinghorne, published in the USA by Trinity Press International, 1991, 36.

[8] I. Prigogine and I. Stengers, *Order Out of Chaos* (London: Heinemann, 1984); see also J. C. Polkinghorne, *Science and Creation* (New Science Library, 1988), ch. 3.

which we have encountered. In an as yet small and imperfect way, one might hope to begin to see some chance of gaining modest insight into how the levels of physics and biology might eventually be found to interlock in their description of the world. Prigogine and Stengers say of their account of these matters that 'we can see ourselves as part of the universe we describe.'[9]

Wonderful! But is it all an illusion? How really open are chaotic systems? Certainly, they are unpredictable, but that is because of the inexactitude of our knowledge of initial conditions, combined with these systems' exquisite sensitivity to the precise character of those conditions.[10]

* * *

As the mathematical physicist reads the situation 'from below,' what will often appear to be happening is mere unpredictability. Out of determinism has arisen apparently random behaviour, but the underlying reality is still held to be purely mechanical. Our limited intellectual powers force us scientifically to think from bottom to top, from underlying simplicity to overall complexity, at least initially. Scientists need a manageable starting-point for their discussions, either in terms of elementary constituents or in terms of a model of abstracted simplicity. We are not clever enough to start with complexity.

A mathematician readily grasps the simple rule defining the Mandelbrot set and then comes upon that set's unlimited richness of structure with great surprise. No one, not even Mandelbrot himself, has the ability to start with the set, to grasp it *ab initio*. Analogously, when we talk about the structure of matter, we start with the simplicities of elementary particle physics rather than the complexities of the theory of condensed matter or of biology.

Our thought is constrained to a one-way reading of the story, in which the higher emerges from the lower. In consequence the latter retains its hold upon our mind as controlling the metaphysical picture. It is by no means clear that this is more than a trick of intellectual perspective. In other words, the characteristics of the elementary level (whether deterministic, or quantum mechanical, or whatever) may be as much emergent properties (in the direction of increasing simplicity) as are life or consciousness (in the direction of increasing complexity). Subatomic

---

[9] Prigogine and Stengers, *Order Out of Chaos*, 300.
[10] The preceding section is from *Reason and Reality* by John Polkinghorne © 1991 by John Polkinghorne, published in the USA by Trinity Press International, 1991, 38.

particles are not only not 'more real' than a bacterial cell; they also have no greater privileged share in determining the nature of reality. That structured chaos can arise from deterministic equations is a mathematical fact. That fact by itself does not settle the metaphysical question of whether the future is determined or, on the contrary, the world is open in its process.

It might, perhaps, be suggested that quantum theory has already settled that issue for us. The most widely held interpretation of that theory's meaning regards individual quantum events as being radically random, so that when the wavefunction 'collapses' on to one of the possible results of a macroscopic observation, the process of the physical world has taken a turn in a particular and intrinsically novel direction.[11] Something unforeseeable has come about. The apparent regularity of so much macroscopic experience is held simply to be the statistical effect of the law of large numbers, the essentially predictable average of many stochastic events.

One might then go on to suppose that, in the case of macroscopic systems in regimes of chaotic behaviour, their exquisite sensitivity to detailed circumstance would effectively enmesh them in a microscopic world of quantum uncertainty. (In attempting prediction one would soon reach levels of required accuracy which are denied to us by Heisenberg's uncertainty principle.) Thus the openness of physical process would seem to have been established, even from a bottom-up point of view. In fact, the matter is more complicated than that, for three reasons.

The first complication relates to the character of quantum physics. If one takes a foundational view of the role of elementary particles, the Schrodinger equation is the true equation, rather than any of those proposed by classical physics. At the time of writing there is a hot debate about whether this equation generates chaotic behaviour. It is certainly known that the analogues of some systems which are classically chaotic (for example, the so-called 'kicked rotator') are not chaotic quantum mechanically. Intuitively one might conjecture that this had something to do with quantum fuzziness on length scales of the Compton wavelength and less, which would not permit the infinitely repeating fractal behaviour which seems to be associated with true chaos.[12] It is not known how

---

[11] See N. Herbert, *Quantum Reality* (New York: Random House, 1985), ch. 8; J. C. Polkinghorne, *The Quantum World* (Longman, 1985), ch. 6.

[12] One might have guessed that it was due to the linearity of the Schrödinger equation, but this does not seem to be the case; see P. C. W. Davies (ed.), *The New Physics* (Cambridge: Cambridge University Press, 1989), 369.

typical are these quantum systems which have been studied and found not to be chaotic. Perhaps quantum mechanics requires a different characterization of chaotic behaviour from that found to work in classical mechanics. It would be extremely perplexing if chaos were totally absent from the quantum world, especially in the limit as Planck's constant becomes small, where correspondence principle arguments encourage the expectation of recovering classically describable behaviour. Joseph Ford has commented that 'Should chaos not be found in quantum mechanics, then an earthquake in the foundations of physics appears inevitable, say about magnitude twenty on the Richter scale.'[13]

A second reason for caution is that the whole question of the nature of quantum reality is itself still a highly contentious issue. Our discussion so far has been in terms of the mainstream understanding held by most physicists (which I personally share). There are, however, radically different proposals which also have their supporters. David Bohm's deterministic version is as empirically adequate as the conventional account, even if it appears to many unpersuasively contrived. The many-worlds interpretation holds that everything that can happen, does happen, even if that implies many alternative yet realized histories for the universe. I am not at all convinced by either of these options, but they remain on the metaphysical table and so they put question marks against any simple claim that quantum theory by itself establishes the openness of physical process.

A third complication relates to an unresolved problem in the interpretation of quantum theory. How does a fitful theory yield a definite observational answer each time it is investigated experimentally? The measurement problem in quantum theory has received no agreed solution but among the possibilities being canvassed is one which would see quantum theory itself as a downward-emergent approximation to a more complex physical reality. The matter is somewhat technical, and certainly contentious.

These considerations lead one to be cautious about invoking quantum theory to establish the openness of macroscopic process. We are encouraged to go on to inquire about the possibility of augmenting bottom-up thinking by intellectual traffic in the opposite direction. Accordingly, I return to the question of whether some of the characteristics discerned in low level exploration of the world (basic physics) may not be regarded

---

[13] J. Ford, 'What is chaos that we should be mindful of it?', in Davies, *The New Physics*, 366. Ford's article contains a good account of the problem of quantum chaos.

as emergent at that level, so that they need not be made universally pre-scriptive for metaphysics. To address the issue bluntly: if apparently open behaviour is associated with underlying apparently deterministic equations, which is to be taken to have the greater ontological seriousness – the behaviour or the equations? Which is the approximation and which is the reality? It is conceivable that apparent determinism emerges at some lower levels without its being a characteristic of reality overall. For instance it might arise from the approximation of treating subsystems as if they were isolatable from the whole, which in fact they are not. But first let us con-sider a philosophical argument.

I take a critically realist view of our scientific exploration of the world. Such a position implies the possibility of gaining verisimilitudinous know-ledge, which is reliable without claiming to be exhaustive. In that case, what we know and what is the case are believed to be closely allied; epis-temology and ontology are intimately connected. One can see how nat-ural this view is for a scientist by considering the early history of quantum theory. Heisenberg's famous discussion of thought experiments, such as the gamma-ray microscope, dealt with what can be measured. It was an epistemological analysis. Yet for the majority of physicists, it led to onto-logical conclusions. They interpret the uncertainty principle as not being merely a principle of ignorance (as Bohm, for example, would interpret it) but as a principle of genuine indeterminacy. In an analogous way, it seems to me a coherent possibility to interpret the undoubted unpredict-ability of so much of physical process as indicating that process to be ontologically open.

The option is there, but it is not, of course, a forced move to choose it. The case for doing so is greatly enhanced if one acknowledges the neces-sity of describing a physical world of which we can see ourselves as inhab-itants. There are, of course, metaphysical traditions which deny that necessity. Cartesian dualism draws a sharp distinction between a realm of pure extension, in which even animals are only automata, and the human realm of minds-in-bodies. I reject that picture and attempt to replace it with a complementary mind/matter metaphysic which sees the world-stuff as being in an emergent-downwards mode the matter of which physics speaks, and in an emergent-upward mode the mind that we experience (the direction being that of increasing complexity and flexibility of organization).[14]

---

[14] Polkinghorne, *Science and Creation*, ch. 5.

There is some relation here with the thought of Jürgen Moltmann, innocent as it is of any detailed concern for scientific insight. In his discussion of what it can mean in the Creed to say that God is the Creator of 'heaven and earth,' Moltmann decides that creation is an open system and 'We call the determined side of this system "earth," the undetermined side "heaven".'[15] One might say that 'earth' is process read downwards towards determinism, 'heaven' is process read upwards towards participation in spiritual reality.[16] A consequence of the delicate sensitivity of complex dynamical systems to circumstance is that they are not only unpredictable but also intrinsically unisolatable.[17]

* * *

The notion of a set of isolated basic entities is a highly abstracted idea. As an elementary particle physicist, I do not question the utility of the notion for some purposes, only its adequacy for all.

That message is reinforced by further consideration of the quantum world itself. I now look to aspects of the subject which are not matters of disputed interpretation, like some of those considered earlier. Whatever one's views on those issues, the theoretical analyses of John Bell and the experimental investigations of Alain Aspect and his collaborators have made it clear that there is an inescapable non-locality involved in the phenomena.[18]

Quantum entities exhibit a counterintuitive togetherness-in-separation, a power once they have interacted to influence each other however far they subsequently separate. Paradoxically, the atomic world is one that cannot be described atomistically. A very careful and lucid discussion of the issues that this raises has been given by Bernard d'Espagnat.[19] He is emphatic that philosophy must take account of what physics has to tell it. 'We may imagine that to reach the truth we only need to come up with brilliant ideas' but that is mistaken for 'it remains illusory to hope that in

---

[15] J. Moltmann, *God in Creation: An Ecological Doctrine of Creation* (London: SCM Press, 1985), 163.

[16] The preceding section is from *Reason and Reality* by John Polkinghorne © 1991 by John Polkinghorne, published in the USA by Trinity Press International, 1991, 39–42.

[17] Ibid., 43.

[18] See, for example, Polkinghorne, *The Quantum World*, ch. 7.

[19] B. d'Espagnat and J. C. Whitehouse, *Reality and the Physicist* (Cambridge: Cambridge University Press, 1989).

our day people can still make valid claims on matters such as reality, time and causality, if these claims are not rooted in the extraordinarily elaborate factual knowledge now at our disposal.'[20] D'Espagnat is a realist, for he feels that denial of an independent reality leads to the danger of collapse into solipsism, a person being driven to retreat into the sole refuge of his own thinking mind. Yet quantum theory denies the possibility of embracing a naive and particulate objectivity in our account of the physical world. D'Espagnat summarizes the dilemma:

> It was once thought [for example by positivism] that the notion of being must be repudiated. Now that it has finally become apparent that to do so is to court incoherence, it is dismaying to find that in the interim it has become peculiarly difficult, if facts are to be respected, to rehabilitate that notion.[21]

His solution is to speak of independent reality as 'veiled' and to be distinguished from empirical reality. That sounds at first like a proposal to move in a Kantian direction of discriminating between phenomena (things as they appear) and noumena (things in themselves), but d'Espagnat does not go all the way with Kant. He insists that independent reality is veiled rather than inaccessible; it is elusive rather than absolutely unknowable. He wishes (as I do too) to give all due weight to the insights of physics but he also acknowledges that 'it does not seem incoherent to me to admit the possibility of rational activity that does not issue in "demonstrative certainty" in the sense we scientists use the expression.'[22] Because I feel very strongly that this is so, I am driven to greater metaphysical boldness than d'Espagnat will permit himself. Nevertheless, I believe that his cautious invocation of veiledness is, at the least, not inconsistent with the kind of openness about the nature of reality which I am trying to explore.

The picture which has been building up is that of a physical world liberated from the thrall of the merely mechanical but retaining those orderly elements which science has been so successful in exhibiting and understanding. In Popper's famous metaphor, it is a world of clouds and clocks, in which some things are indeed predictable but others are open to the possibility of new development. I have elsewhere argued that such a world of intertwined order and novelty is just that which might be

---

[20] Ibid., 16.
[21] Ibid., 11.
[22] Ibid., 210.

expected as the creation of a God both faithful and loving, who will endow his world with the twin gifts of reliability and freedom.[23]

In a bottom-up description of the physical world, the onset of flexible openness is signaled by the myriad possibilities of future development which present themselves to a complex dynamical system. In a quasi-determinist account they arise from the greatly differing trajectories which would result from initial conditions differing only infinitesimally from each other. Because of their undifferentiable proximity of circumstance, there is no energetic discrimination between these possibilities. The 'choice' of path actually followed corresponds, not to the result of some physically causal act (in the sense of an energy input) but rather to a 'selection' from options (in the sense of an information input).

One might well be able to formalize the last point. Typically the open options can be expressed in terms of bifurcating possibilities (this or that), whose particular realizations resemble bits of information (switches on or off, in a crude computer analogy). In a top-down description of systems of such complexity as ourselves, this 'information input' is a picture of how mind could operate causally within a complementary mind/matter metaphysic. Because flexibility only arises within intrinsically unpredictable circumstances, the springs of the operation of mind would be inescapably hidden ('veiled'). The search for a modern equivalent of the Cartesian pineal gland would be the search for a will-o'-the-wisp; it is condemned to failure.

It is by no means clear that information input of the kind described originates solely from animals, humankind, and whatever similar agents there might be. I do not believe that God is contained within the mind/matter confines of the world,[24] but it is entirely conceivable that he might interact with it (both in relation to humanity and in relation to all other open process) in the form of information input. I have attempted elsewhere to explore some of the theological consequences of such a view, particularly in relation to questions of prayer and theodicy.[25] God is not pictured as an interfering agent among other agencies. (That would correspond to energy input.) Instead, form is given to the possibility that he influences his creation in a non-energetic way.

Many theological writers have recoiled from the detachment of deism and have wished to assert an interactive relationship between God and

---

[23] Polkinghorne, *Science and Creation*, ch. 4.
[24] See ibid., 79–82.
[25] J. C. Polkinghorne, *Science and Providence* (New Science Library, 1989).

the world. They have been notably coy, however, about how this might actually come about. Austin Farrer's account of double agency is so emphatic about the inscrutability of the divine side of it as to provide us with no help.[26] The various varieties of panentheism (asserting the world to be part of God, but not the whole of him) afford no more than an image of divine action – and an unsatisfactory one at that, in my opinion.[27] Arthur Peacocke has offered us the picture of God as 'an Improviser of unsurpassed ingenuity',[28] seeking to incorporate the discords of evil into a greater harmony. How that Great Improviser actually touches the keyboard is not made clear. The idea of divine interaction through information input[29] seems to me to afford us some help in the matter.[30]

[26] See ibid., 11–13.

[27] Ibid., ch. 2.

[28] Peacocke, *God and the New Biology*, 98.

[29] There are connections here with Bowker's notion of religions as systems; see J. Bowker, *Licensed Insanities: Religions and Belief in God in the Contemporary World* (London: Darton, Longman & Todd/Meakin and Associates, 1987), 112–43.

[30] The preceding section is from *Reason and Reality* by John Polkinghorne © 1991 by John Polkinghorne, published in the USA by Trinity Press International, 1991, 43–6.

3

Human nature

No scientist has had greater influence on general human thinking than Charles Darwin. The publication of the *Origin of Species* in 1859 was a climactic point in a process started by the geologists towards the end of the eighteenth century, which recognized the inherent changeableness of the natural world and the consequent necessity of an historical mode of thinking if one is to gain a proper understanding of its character. It is impossible for us today to consider the present without seeking to take into account its origin in the past. Time is no longer simply the index of when events happened, but it signifies and contains the evolutionary process through which things have come to be.

A world in which species were stable, totally immune from change, might well have been capable of being thought of theologically as a creation that had sprung into being readymade, its origin simply the result of the direct action of the God who was that world's Designer. On the other hand, a world of radical temporality, in which change is the engine driving the emergence of novelty, is one to which its Creator's relationship has to be understood in somewhat different terms. In words used by both Charles Kingsley and Frederick Temple in the aftermath of the publication of the *Origin*, an evolving world may appropriately be thought of theologically as a creation in which creatures are 'allowed to make themselves.' In other words, from a theological perspective, evolution is simply the way in which creatures are allowed to explore and bring to birth the fruitfulness with which the Creator has endowed creation.

When the initial impact of Darwin's great discovery had somewhat abated, other theologians followed in the wake of Kingsley and Temple in welcoming these new insights. Many came to see that evolutionary thinking was not incompatible with a doctrine of creation, but the latter would have to be expressed somewhat differently from the way that had been customary in the past. Emphasis now needed to be laid not only on *creatio ex nihilo* (understood as asserting the will of God to be the cause not only of how things began but also why things remain in existence), but also on a complementary process of unfolding *creatio continua*. With the benefit

36

of hindsight, this story of development might even be thought, anachronistically, to have been hinted at in the sequence of the 'days' of creation in Genesis 1, and by Augustine's notion that the initiating timeless act of creation brought into being the 'seeds' from which eventually a multiplicity of different creatures would develop in the course of a process of temporal germination.

Such ideas are not at all uncongenial to a theology that sees God as the ordainer of the character of creation, and so as the One who is to be thought of as acting as much through natural processes as by any other means. Unfolding evolution simply expresses the divine intention for the way in which creation is to realise its God-given potentialities. The contingencies present in the process represent the Creator's gift to creatures of the freedom to make themselves. To suppose the contrary, and to posit an opposition between natural process and divine purpose, would be to fall into the Manichean heresy, the supposition that God and the world are totally at odds with each other.

Much more shocking to some theologians, and to much human sensibility generally, was the idea, implicit in the *Origin* but only later stated clearly by Darwin (who was always somewhat apprehensive about public reaction to what he had to say, so that he waited until the publication of *The Descent of Man* in 1871 before he made the point explicitly) that evolutionary thinking must embrace human origin as well as that of the other creatures. The consequent implication of humanity's close degree of kinship with the animals was by no means a welcome thought to all. Many, whether religious believers or not, felt that it carried the threat of an intolerable subversion of human dignity and status. This fear underlies the unedifying exchange about monkey ancestry said to have taken place between Bishop Samuel Wilberforce and 'Darwin's bulldog,' Thomas Henry Huxley, in their notorious encounter at the British Association meeting at Oxford in 1860.[1]

It is impossible for us today to think about human nature without acknowledging the significance of the evolutionary origin of *Homo sapiens*. While this fact certainly implies a degree of cousinly relationship between humankind and the other animals (particularly the higher primates), it by no means implies that this recognition exhausts all that needs to be said, as if human beings are just another kind of ape.

---

[1] There seems to be some historical uncertainty about what actually happened; see J. H. Brooke, *Science and Religion* (Cambridge: Cambridge University Press, 1991), 40–2.

The history of the development of life has been characterized by a sequence of unprecedented emergences, leading to the appearance on the terrestrial scene of that which is qualitatively novel. The first such emergence was life itself, when the complexifying organisation of inanimate matter reached a stage that produced systems that could maintain and reproduce themselves. After about two billion years during which all living entities were single-celled, there emerged the increasing complexity of multicellular organisms. Eventually this led to the dawning of consciousness for the first time on planet Earth.

In turn, we should not hesitate to recognise that there are a number of characteristics of human nature, and perhaps of the natures of our immediate hominid ancestors, that clearly mark out the genus *Homo* as constituting a yet further emergent level of fundamental and astonishing novelty. Of course, we have evolved from previous forms of animal life, but one should not commit the genetic fallacy of supposing that knowledge of origin is the same as knowledge of nature. I want briefly to survey a number of characteristics that support a claim for unique human status.

The first point is that humans are *self-conscious beings* in a radically new way. Of course, the higher animals are conscious, but they seem to live in what we may call the near present. The chimpanzee can foresee that throwing the stick may dislodge the banana, and understand that the person who just now went behind the rock is still there though hidden from sight. However, the power to look far ahead, even to the point of brooding on the thought of our eventual deaths, despite the fact that they may lie many years in the future, seems to be a capacity that only humans possess. It is said that elephants show signs of sorrow at the death of one of their number, but again this is something related only to the near present. Part of humanity's unique self-consciousness is a keen awareness of ourselves, not just as recognised when we see our images reflected in a mirror, but as persons whose characters are formed by our experiences of life, reflectively and reflexively assessed.

The human possession of *language*, with its profound conceptual range and almost limitless flexibility to respond to novel experiences and changing situations, is clearly linked with the human exercise of self-conscious reflection. It enables that very characteristic human activity of story-telling, and the remarkable possibility of writing poetry. As with a number of other human characteristics, one can see some pre-figuration of linguistic abilities in the capacities possessed by the higher apes and some cetaceans.[2] However, the differences in degree between them and human beings remain so great as to amount to a qualitative distinction. It is certainly interesting that prolonged human intervention can induce in selected chimpanzees the ability to manipulate a limited vocabulary of signs and to form simple 'sentences,' but these achievements fall very far short indeed of the creativity of human linguistic attainments.

More generally, human beings possess a great range of *rational skills*. Later we shall pay some attention to mathematics, but for the moment one may simply illustrate the fruitfulness of human reasoning by appealing to the astonishing extent of our scientific understanding. With the dawning of hominid self-consciousness not only did the universe become aware of itself, but a process began through which the secrets of its structure and history would progressively come to be unveiled. Even such counterintuitive regimes as the subatomic world of quantum theory, or the vast expanses of cosmic curved spacetime, radically different in their character from the world of everyday experience and remote from direct impact on it, have proved to be open to human enquiry and understanding. Human ethologists write substantial volumes analysing the behaviour of communities of chimpanzees, thereby displaying an insightful interest in their character, but this is not reciprocated by the primates in their turn. Learning how to use a stone to crack a nut is a valuable skill, but it scarcely bears comparison with the remarkable devices produced through human technological invention.

Human beings also possess great *creative powers*, manifested through art and culture. From the time of the earliest cave paintings known to us, there is evidence of an engagement with the aesthetic, displaying a quality that defies explanation in merely pragmatic terms. The beautiful notes of birdsong are apparently principally a means of asserting territorial possession, but humans explore the inexhaustible riches of music for reasons that centre on delight rather than utility.

We are also *moral beings* in a way that does not seem to be the case for the animals, for whom concepts of right and wrong and of ethical obligation do not appear to be appropriate. When we read medieval stories of an ox being put 'on trial' for goring a man to death, and then 'executed' for the misdeed, they seem to us to be grotesque, the result of a foolish kind of category mistake. I am sure that we should treat animals with ethical respect, but grounding this attitude in a doctrine of animal rights

---

[2] See the discussion of the communication skills of dolphins in S. Conway-Morris, *Life's Solution* (Cambridge: Cambridge University Press, 2003), 250–3.

seems to me to be mistaken, for the use of rights language would surely require there to be a complementary doctrine of animal duties, a notion that does not appeal to us as being reasonable.[3]

At almost all times and in almost all places, human beings have participated in an admittedly bafflingly diverse history of encounters with the sacred, experiences that indicate that we possess a capacity for what may fittingly be called *God-consciousness*. Mystical apprehension of unity with the One or the All; numinous encounter with the mysterious and fascinating reality of the divine standing over and against humanity in mercy and judgment; the less dramatic engagements of regular worship; all these are human experiences of great intensity which many of us value as being of undeniable authenticity and significance, but which have no discernible counterpart in the lives of our animal cousins.

Finally, theologians detect in human life a slantedness which they categorise as *sin*, a source of distortion in human affairs that frustrates hopes and corrupts intentions. Of course, among the animals there is not only necessary predation, but also what appears to be a kind of cruel mischief, as when apes sometimes wantonly tear apart small monkeys whom they have happened to come upon. Of course, among the animals there can be fierce struggles over access to food or mating, or a challenge to the hegemony of the α-male, though these are not often fought literally to the death.

In contrast, there appears to be in human history, with its wars, crusades and acts of genocide, a degree of depravity that seems to be on an entirely different scale from these animal incidents. And these communally willed human acts of immense evil are the shadows writ large of the many lesser acts of betrayal and selfishness that occur all the time in the individual circumstances of everyday life. We are not only moral beings, but through our actions we show ourselves to be flawed moral beings. Reinhold Niebuhr once observed that the only empirically verifiable Christian doctrine is that of original sin. Simply look around you, or within your own heart, and you will see that it is true. Theologians diagnose the root of sin as lying in human alienation from the God whose gift of the spiritual power that theologians call grace is the proper ground of a fully human life, a theme to which we shall return.

Even so brief a survey indicates how strange it is that many biologists can claim not to be able to see anything really distinctive about *Homo*

---

[3] Of course, human neonates have rights but not yet duties, but they will grow into an ethical awareness that will bring with it moral responsibility. We do not see signs of a corresponding ethical development in animals.

sapiens. They regard human behaviour as just another instance of animal behaviour, and humanity as a not particularly special twig on the burgeoning bush of evolutionary development. In fact, even the ability to articulate these assertions is sufficient to deny their premise, providing as it does an example of humankind's unique capacity for self-reflexive thought. The fact that we share 98.4 percent of our DNA with chimpanzees shows the fallacy of genetic reductionism, rather than proving that we are only apes who are slightly different. After all, I share 99.9 percent of my DNA with Johann Sebastian Bach, but that fact carries no implication of a close correspondence between our musical abilities.

I must confess to being a speciesist – provided that term is understood as involving an acknowledgement of the true novelty resulting from hominid emergence, but not if it is taken as implying a failure to accord the proper kind of respect to the animals. Yet, the ethical basis for this respect takes a different form from that which underlies the conviction that human persons are each of unique value, so that they are never to be treated as means but only as ends. Many of us consider that it is permissible in a humane way to cull a herd of deer facing a severe shortage of fodder, thereby preserving the type at the expense of the deaths of some of the individuals. Such a strategy in relation to a human community faced with famine would universally be acknowledged to be unethical.

Our concern should be to take absolutely seriously the fact that human beings have emerged through a long and continuous history of biological evolution, while at the same time taking with equal seriousness the fact of the qualitative differences that correspond to the irreducible novelty of that emergence. The first consequence of acknowledging human evolutionary origin is to reinforce an understanding of human beings as *psychosomatic unities*. A dualist picture of humans as possessing a spiritual soul encased in a material body from which it is potentially separable is not absolutely ruled out. One could conceive of the Creator as bestowing an extra dimension of spiritual being on a purely material entity that had attained the degree of complex development that would make this gift appropriate (something like this appears to be the official Vatican view). Yet it is surely much more persuasive to think of humans as animated bodies, a kind of 'package deal' of the material and the mental and spiritual in the form of a complementary and inseparable relationship. The aim is to take human embodiment absolutely seriously, without falling into the error of treating our mental and spiritual experiences as no more than epiphenomenal fringe effects of the material. Such an integrated understanding of humanity would not have surprised the writers of the

41

Bible, for this was the predominant way in which Hebrew thought regarded human nature.

Taking a psychosomatic view of human beings requires careful thinking about how to understand the nature of the human *soul*.[4] The concept is one of central importance to theology, where the soul has the role of being the carrier of the intrinsic essence of individual personhood – the 'real me,' as one might say. Theological anthropology cannot abandon this use of soul language, though it has to be able to free itself from bondage to a past heritage of conceiving it in terms of platonic categories.

The 'real me' is certainly not to be identified merely with the matter of my body, for that is continually changing through the effects of wear and tear, eating and drinking. What maintains continuity in the course of this state of atomic flux is the almost infinitely complex *information-bearing pattern* in which the matter of the body is at any one moment organized. It is this pattern that is the human soul. The idea has a venerable history, for both Aristotle and Thomas Aquinas thought of the soul as being the 'form' of the body, its constitutive and organising principle, so to speak – a concept that obviously has a close connection with the position being advocated here. Three things must be said about this modern version of the way of conceiving of the soul.

The first point is that the flat language used in the phrase 'information-bearing pattern' is an almost wholly inadequate attempt to point to a necessarily much richer concept, lying beyond anything that we are presently able to articulate properly. Human persons are relational beings and the patterns that constitute them cannot simply end at their skin. The rich variety of human capacities and forms of experience, briefly surveyed in the course of the defence of human uniqueness, must also find appropriate incorporation into the nature of the human soul. The concept of information must be enriched sufficiently to accommodate the mental and spiritual dimensions of human nature. Aristotle and Aquinas thought of the soul as being the principle of life animating a body, and they believed that there are vegetative and animal souls as well as human souls. In terms of the present discussion, what differentiates the human soul would be precisely the rich, many-layered complexity that is a reflection of the unique range of human capacities.

The second point is that the picture I am proposing is a dynamic one, for the pattern that is my soul will develop as my character forms and my

---

[4] J. C. Polkinghorne, *The God of Hope and the End of the World* (SPCK/Yale University Press, 2002), 103–7.

experiences, understandings and decisions mould the kind of person that I am. John Keats's image, in one of his letters to Fanny Brawne, of this life as a vale of soul-making is an apposite one. Understood in this way, the soul is not a once-for-all gift, as if it were fully conveyed at conception or at birth, but it has its individual history. This observation does not rule out there being an unchanging component in the soul, which could be thought of as the signature identifying a specific person, but this would only be a part of what makes it up. (The individual genome would, presumably, be a constituent of this invariant sub-pattern.)

The third point is to recognise that this way of understanding the soul implies that it does not of itself possess an intrinsic immortality. As far as a purely naturalistic account could take us, the information-bearing pattern carried by the body would be expected to dissolve at death with that body's decay. It is, however, a perfectly coherent possibility to deepen the discussion by adding a theological dimension, and to affirm the belief that the God who is everlastingly faithful will preserve the soul's pattern *post mortem* (holding it in the divine memory is a natural image), with the intention of reconstituting its embodiment in a new environment through the great divine eschatological act of the resurrection of the whole person. This way of formulating the Christian hope of a destiny beyond death, which would have been perfectly acceptable to Aquinas, is one that I have explored elsewhere.[5] Its basic structure is the pattern of death and resurrection, rather than the notion of spiritual survival.

It is important to recognise that evolutionary thinking is as much concerned with environment as it is with genetics. This point has been emphasised particularly by Holmes Rolston in his reflections on biological process.[6] It was Darwin's great genius to understand how interaction between individual variations (only later recognised as stemming from genetic mutations) and the constraint of survival in a specific ecological setting could together, through the shuffling exploration and sifting of small differences, so shape the development of living forms over long periods of time as to induce a fruitful history of increasing complexification.

Orthodox neo-Darwinian thinking conceives of the environment relevant to evolution, including that of humanity, as being one that can be understood solely in physico-biological terms, and the mode of interaction with it as being solely the process of differential reproductive success.

---

[5] Ibid., 107–12.

[6] H. Rolston, *Genes, Genesis and God* (Cambridge: Cambridge University Press, 1998).

While we can agree that natural selection has been an important factor in the development of life on Earth, it is by no means obvious that it is the only type of process involved.[7] The timescale of the history of terrestrial life is certainly long (3.5 to 4 billion years), but the train of developments that has to be accommodated within that span is immensely complex, with the first two billion years or so being taken up solely by single-celled organisms.

Argument on this general point is not, however, our present concern. Instead, our focus is on the emergence of hominid life, taking place over a period of just a few million years and with modern *Homo sapiens* only appearing in the last one to two hundred thousand years. The central issue is whether a strictly neo-Darwinian account, with its narrow concept of the effective environment within which these developments took place, is sufficient to explain the coming-to-be of the many distinctive features that we have claimed mark off human nature from other forms of animal life.

In fact, the attempt to force classical Darwinian thinking into the role of an explanatory principle of almost universal scope[8] has proved singularly unconvincing as it seeks to inflate an assembly of half-truths into a theory of everything. Sober evaluation of the adequacy of the insights being proffered soon pricks this explanatory bubble. Increasing ability to process information coming from the environment is clearly an advantage in the struggle for survival, but this does not explain why it has been accompanied by the property of conscious awareness. Indeed, one might suppose that the latter, with its limited focus of attention and no more than a peripheral openness to signals coming from other possible directions, might be more a hazard than a help.

Evolutionary epistemology[9] has attempted to explain and validate human rational powers to attain reliable knowledge as being something originating through Darwinian development. Once again, one encounters

---

[7] For discussion of the role of the self-organising properties of complex systems, see B. Goodwin, *How the Leopard Changed Its Spots* (London: Weidenfeld and Nicolson, 1994; S. Kauffman, *The Origins of Order* (Oxford: Oxford University Press, 1993). See also the evidence for the convergence of different genetic lines onto constrained possibilities for biologically accessible and functionally useful structures, discussed in Conway-Morris, *Life's Solution*.

[8] D. Dennett, *Darwin's Dangerous Idea* (New York/London: Simon and Schuster, 1995); E. O. Wilson, *Consilience* (New York: Knopf, 1998).

[9] P. Munz, *Our Knowledge of the Growth of Knowledge* (London: Routledge & Kegan Paul, 1985); F. Wuketits, *Evolutionary Epistemology* (New York: State University of New York Press, 1990); W. van Huyssteen, *Duet or Duel?* (London: SCM Press, 1998).

a half-truth. Of course, being able to make sense of everyday experience is a vital survival asset. If one could not figure out that stepping off a high cliff was a dangerous thing to do, life would soon be imperiled. Yet when Isaac Newton recognised that the same force that makes the high cliff dangerous is also the force that holds the Moon in its orbit around the Earth and the Earth in its orbit around the Sun, thereby going on to discover universal gravity, something happened that went far beyond anything needed for survival.

When Sherlock Holmes and Dr Watson first meet, the great investigator feigns not to know whether the Earth goes round the Sun or the Sun around the Earth. He defends his apparent ignorance simply by asking what it matters for his daily work as a detective. Of course, it does not matter at all, but human beings know many things that neither bear relation to mundane necessity nor could plausibly be considered simply as spin-offs from the exercise of rational skills developed to cope with those necessities. This point was reinforced about two hundred years after Newton when Albert Einstein's discovery of general relativity produced the modern theory of gravity, capable of explaining not only the behaviour of our little local solar system but also the structure of the whole cosmos. In both relativity theory and quantum theory, modes of thought are required that are totally different from those appropriate to everyday affairs. Yet, as we have already noted, the human mind has proved capable of comprehending the counterintuitive world of subatomic physics and the cosmic realms of curved spacetime.

It has turned out that it is our mathematical abilities that have furnished the key to unlock deep secrets of the physical universe. Once more one encounters a mystery impenetrable to the conventional evolutionary thinking. Survival needs would seem to require no more than a little arithmetic, some elementary Euclidean geometry, and the ability to make certain kinds of simple logical association. Whence then comes the human ability to explore non-commutative algebras, prove Fermat's Last Theorem, and discover the Mandelbrot set? These rational feats go far beyond anything susceptible to Darwinian explanation.

The ability to use the experience of yesterday as a guide to coping with the challenges of today is clearly a significant aid to survival. But does this fact alone give us sufficient licence to trust in human ability to reconstruct from fragmentary evidence the history of a past extending over many millions of years? Darwin himself felt some doubts on this score, writing in old age to a friend, 'With me the horrid doubt arises whether the convictions of man's mind, which has been developed from the mind of lower animals, are of any value or are trustworthy.'[10] There is something touching in this spectacle of this great scientist poised with rational saw in hand and tempted to sever the epistemic branch on which he had sat while making his great discoveries. Surely his doubts were unjustified. The cumulative power of scientific thinking has vindicated itself many times over in the course of human investigations into reality. Why science is possible in this deep way is a question which, if pursued, would take us well beyond science itself.[11] I believe that ultimately it is a reflection of the theological fact that human beings are creatures who are made in the image of their Creator (Genesis 1.26–27).

Sociobiology[12] seeks to explain human ethical intuitions in terms of inherited patterns of behaviour favouring the propagation of at least some of an individual's genes. Once again, one may acknowledge a source of partial insight. No doubt ideas of kin altruism (the mutual support extended between those who share in the family gene pool) and reciprocal altruism (favours done in the expectation of favours later to be received) shed some Darwinian light on aspects of human behaviour. Games theoretic models of behavioural strategies that optimise probable returns in given circumstances – such as 'tit for tat': respond in the same manner that your opponent has displayed to you – give some insight into the nature of prudent decision making. But sociobiology tells too banal a story to be able to account for radical altruism, the ethical imperative that leads a person to risk their own life in the attempt to save an unknown and unrelated stranger from the danger of death. Love of that incalculable kind eludes Darwinian explanation.

Equally elusive to evolutionary explanation are many human aesthetic experiences. What survival value has Mozart's music given us, however profoundly it enriches our lives in other ways?

The proper response to all this is not to adopt a Procrustean technique of chopping down the range of human experience until it fits into a narrow Darwinian bed, nor is it to abandon evolutionary thinking altogether. Rather, it is to release that thinking from the poverty of its neo-Darwinian captivity. This requires two steps. One is an enrichment of the concept of the environment within which hominid evolution has taken place. The other is an enhancement of our understanding of the

---

[10] Quoted in M. Ruse, *Can a Darwinian Be a Christian?* (Cambridge: Cambridge University Press, 2001), 107.

[11] See J. C. Polkinghorne, *Beyond Science* (Cambridge University Press, 1996).

[12] E. O. Wilson, *Sociobiology* (Cambridge, Mass.: Belknap Press, 1975).

processes that have been at work. When these steps have been taken, we shall be freed from being driven to the construction of implausible Just-so stories, alleging that human capacities of which we have basic experience are totally different in character from what we, in fact, know them to be.

One way of enhancing understanding of the actual scope and character of the human environment can be provided by thinking about the nature of mathematics. Most mathematicians are convinced that their subject is concerned with discovery and not with mere construction.[13] They are not involved in playing amusing intellectual games of their own contrivance, but they are privileged to explore an already-existing realm of mentally accessed reality. In other words, as far as their subject is concerned mathematicians are instinctive Platonists. They believe that the object of their study is an everlasting noetic world which contains the rationally beautiful structures that they investigate. Benoit Mandelbrot did not invent his celebrated set; he discovered it. Acknowledgement of the existence of this rational dimension of reality is vital to the possibility of understanding the origin of human mathematical powers.

At some stage of hominid development, our ancestors acquired a brain structure that afforded them access to the mental world of mathematics. It then became as much a part of their environment as were the grasslands over which they roamed. At first, this noetic encounter must have been limited to a utilitarian style of mundane thinking, involving just an engagement with simple arithmetical and geometrical ideas. However, once that intellectual traffic had started it could not be limited to such elementary matters. Our ancestors were beguiled into further exploration of noetic richness which, once begun, continued with an ever-increasing fruitfulness. What drew them into this exploratory process was not a Darwinian drive to the enhanced propagation of their genes, but an entirely different mechanism that we shall consider shortly.

The kind of considerations in the case of mathematical experience that lead us to take seriously an enriched human environment apply equally to other distinctive forms of human ability. Human ethical intuitions indicate the existence of a moral dimension of reality open to our exploration. Our conviction that torturing children is wrong is not some curiously veiled strategy for successful genetic propagation, nor is it merely a convention adopted arbitrarily by our society. It is a *fact* about the nature

---

[13] See J. C. Polkinghorne, *Belief in God in an Age of Science* (Yale University Press, 1998), ch. 7.

of reality to which our ancestors gained access at some stage of hominid development. Similarly, human aesthetic experiences gain their authenticity and value from their being encounters with yet another aspect of the multidimensional reality that encompasses humanity. Experiences of beauty are much more than emotion recalled in tranquility; they are engagements with the everlasting truth of being.

Once one accepts the enrichment beyond the merely material of the context within which human life is lived, one is no longer restricted to the notion of Darwinian survival necessity as providing the sole engine driving hominid development. In these noetic realms of rational skill, moral imperative and aesthetic delight – of encounter with the true, the good and the beautiful – other forces are at work to draw out and enhance distinctive human potentialities. Survival is replaced by something that one may call *satisfaction*, the deep contentment of understanding and the joyful delight that draws on enquirers and elicits the growth of their capacities. No doubt the neural ground for the possibility of psychosomatic beings like ourselves to be able to develop aptitudes in this way was afforded by the plasticity of the hominid brain. Much of the vast web of neural networking within our skulls is not genetically predetermined, but it grows epigenetically, in response to learning experiences. It is formed by our actual encounters with reality.

The era of these developments was the time when human culture emerged, generating a language-based Lamarckian ability to transfer information from one generation to the next through a process whose efficiency vastly exceeded the slow and uncertain Darwinian method of differential propagation. It is in these ways that recognition of the many-layered character of reality, and the variety of modes of response to it, make intelligible the rapid development of the remarkable distinctiveness of human nature.

Hominid evolution inaugurated the exercise of these new creaturely capacities here on planet Earth, but it did not bring into being the reality to which these nascent abilities gave access. What emerged were mathematicians, not mathematics. The latter was always 'there,' even if unrecognised by creatures during billions of years of cosmic history. The rational, moral and aesthetic contexts within which hominid capacities began to develop are essential and abiding dimensions of created reality.[14] From the point

---

[14] There may be a contrast here with the ideas of 'emergentist monism.' Philip Clayton says that 'unlike dual-aspect monism, which argues that the mental and the physical are two different ways to characterise one "stuff," emergent monism conceives the relationship between them

of view of dual-aspect monism, these realities exist at the extreme mental pole of the complementary duality involved, just as sticks and stones exist at the far physical pole.[15]

One should go on to ask about the origin of these many diverse but interrelated aspects of reality. For the religious believer, the source of these dimensions lies in the unifying will of the Creator, a fundamental insight that makes it intelligible not only that the universe is transparent to our scientific enquiry, but also that it is the arena of moral decision and the carrier of beauty. Those dimensions of reality, the understanding of whose character lies beyond the narrow explanatory horizon of natural science, are not epiphenomenal froth on the surface of fundamentally material world, but they are gifts expressive of the nature of this world's Creator. Thus moral insights are intuitions of God's good and perfect will, and aesthetic delight is a sharing in the Creator's joy in creation, just as the wonderful cosmic order discovered by science is truly a reflection of the Mind of God. Thinking about human experience in this way affords the possibility of a satisfyingly unified account of multilayered reality. Theology can lay just claim to be the true Theory of Everything.

For the theologian, the most important context within which hominid development has taken place is the veiled but grace-giving presence of God. Not only does this ultimate creaturely setting *coram deo*, before the divine, provide the explanation of humanity's continuing encounter with the sacred, but it also gives a distinctive character to theological anthropology. In contrast to many secular accounts of human nature, a Christian understanding of humankind acknowledges our heteronomous dependence on the grace of God, rather than asserting an autonomous human independence. It is the refusal to acknowledge this status of creaturely dependence that is seen as being the root cause of humanity's sinful predicament. Going it alone, 'Doing It My Way,' is not the prescription for a truly fulfilled life. Human beings are intrinsically heteronomous. Recently, however,

---

as temporal and hierarchical' (in T. Peters and G. Bennett (eds), *Bridging Science and Religion* (London: SCM Press, 2002), p. 116). It is not altogether clear whether this remark is meant to apply simply to the emergence of perceiving entities (in which case there is no conflict with the view taken here), or whether it applies also to what is perceived. The latter notion of emergence would seem to require a constructivist, rather than a realist, conception of the nature of mentally accessed entities. The dual-aspect monism that I espouse seeks to regard the mental and the physical as corresponding to encounters with complementary phases of the 'one stuff' of created reality, rather than simply different characterizations of it.

[15] J. C. Polkinghorne, *Faith, Science and Understanding* (SPCK/Yale University Press, 2000), 95–9.

feminist theologians have offered some trenchant criticisms of this concept of human dependency.[16] Clearly there is a danger that heteronomy, wrongly construed, may lead to the picture of a passive and subservient humanity, of the kind that Nietzsche satirised in his famous comment that Christianity is a religion for slaves. No one could deny that this distortion has from time to time been present in the life of the Church, particularly in relation to the subordination of women. Yet, at its best, Christianity has sought not to fall into this error. The long history of theological debate about the correct balance between the roles of grace and freewill is one aspect of the search for a right understanding of heteronomy. Paul gave classic expression to the matter when he exhorted the Philippians (2.12–13) to 'work out your own salvation with fear and trembling; for it is God who is at work in you, enabling you both to will and to work for his good pleasure.' The words of a collect from the Book of Common Prayer, which speaks of God as the one 'whose service is perfect freedom,' are another fine expression of a dependence upon the Creator which is the ground of human dignity and worth, rather than the threat of their subversion.[17]

---

[16] For a temperate and helpful account of the issues, see S. Coakley, *Powers and Submissions* (Malden, Mass.: Blackwell, 2002).

[17] The preceding section is from *Exploring Reality* by John Polkinghorne © 2005 by Yale University, published in the USA by Yale University Press, reprinted by permission of Yale University Press, 38–59.

4

The nature of reality

I am not content just to say that, in my scientific life, I am concerned with the material universe, while in my ministry in the Church, it is spiritual values that are the objects of my concern – and never mind how they relate to one another. 'No one can serve two masters.'[1] One's instinct to seek a unified view of reality is theologically underwritten by belief in the Creator who is the single ground of all that is. The rich complexity of creation demands an account of the world that will not deny proper respect to the nature of either the mental or the material. It must accommodate within its metaphysical embrace both the constituent insights of elementary particle physics and also the integrating insights of aesthetic and religious experience. It must do justice to human embedding in time, and to man's worshipful intuitions of eternity. Only by attempting to locate thought within such a comprehensive framework can we feel comfortable in considering those particular points of interaction between science and theology which have been our principal concern so far. At the very least, we must have a tentative view of reality that holds together, in a single account, the varied subjects of our discourse.

To state the goal of our inquiry is by no means to imply that it will easily be attained. In fact, of course, if I were able to set before you such a universal understanding, I would have achieved for myself an imperishable place in the history of thought. The actuality is rather different. The proportion of perplexity to insight will be considerable, but for all that I do not feel unduly apologetic. The practice of science encourages one to think that it is better to hold together various puzzling pieces of experience, with respect for their stubborn facticity, and then make whatever shift one can to reconcile them, rather than to produce a tidy scheme by willful oversimplification and neglect of evidence. The early investigators of quantum phenomena had to hang on by the skin of their intellectual

---

[1] Matthew 6.24.

teeth until eventually they found the harmonious resolution of threatened paradox.[2]

The nineteenth-century physicist contemplated a dual world of particles and waves. Newton told him how particles behaved; Maxwell laid down the equations governing the wave-like behavior of the electromagnetic field. Both sorts of entity were necessary, since Young's diffraction experiments had disproved the suggestion, somewhat cautiously advanced by Newton, that light might be made up of tiny particles. The discoveries of Planck and Einstein set in train a line of research that revealed that these apparently distinct possibilities interpenetrated each other. Light ejected electrons from metals in a way that could be made sense of only by adopting Einstein's interpretation that it was behaving like a beam of particles, each endowed with Planck's quantum of energy. Electrons did what de Broglie had suggested that they might do and were diffracted by a thin metal foil in a wave-like way.

Eventually an understanding was achieved that saw wave and particle as complementary aspects of a single reality. Treat a quantum entity as a wave and it will oblige you with the appropriate behavior; treat it like a particle and that will be the way in which it will be found to respond. Such seemingly fickle behavior is nevertheless logically consistent because interrogation in the wave mode and interrogation in the particle mode are mutually exclusive experimental possibilities. You can employ one or the other but not both at once, so that contradiction is never encountered. Complementarity, as the quantum physicists call this delicate behavior, is the scientist's equivalent of the theologian's *perichorēsis*, the mutual indwelling of characteristics.[3]

* * *

The mental and the material are both to be given their due, but that desirable end will not be achieved satisfactorily by espousing a Cartesian dualism, simply setting them side by side and invoking the pineal gland, or divinely arranged synchronization, to explain the manifest correspondences between them. The act of the willed lifting of a hammer and its consequent use to render someone unconscious by a smart tap on the

---

[2] J. C. Polkinghorne, *The Quantum World* (Longman, 1984), ch. 2.
[3] The preceding section is from *Science and Creation* by John Polkinghorne © 1988 by John Polkinghorne, revised edition published by Templeton Press, 2006, 83–5.

head make the existence of such correspondences only too clear. Cartesianism simply fails to explain how they come about in any way that carries conviction. Its twin worlds of extension and thought, matter and mind, fail to coalesce into the one world of our psychosomatic experience.

If it is neither mind nor matter nor mind-and-matter, what remains? The only possibility appears to be a complementary world of mind/matter in which these polar opposites cohere as contrasting aspects of the world-stuff, encountered in greater or lesser states of organization. If you take me apart, you will find that all you get will be matter – in all the elusive subtlety that quantum mechanics has taught us to attribute to the material – matter ultimately found to be constituted of the quarks, gluons and electrons that compose all the rest of the physical universe. Neither soul nor entelechy will be found as a separate part of the residue. Yet if you want to encounter *me*, you will have to refrain from that act of decomposition and accept me in my complex and delicately organized totality. That almost infinitely complex information-carrying pattern, which persists through all the changes of material constituents as nutrition and wear-and-tear ceaselessly replace the individual atoms of my body, and which by its very persistence expresses the true continuity of my person – that pattern is the meaning of the soul.

It seems to me that this understanding corresponds closely to the Aristotelian notion of the soul as the 'form' of the body. St Thomas Aquinas felt that, while this would do very well for understanding the nature of animals, more was needed to do justice to humanity. 'Man,' he wrote 'is non-material in respect of his intellectual power because the power of understanding is not the power of an organ.'[4] I do not see why that should be so. Understanding, self-consciousness, and the ability to know and worship the Creator may all be powers of that wonderfully subtle organ, the brain – or, better still, of the whole body – in striking illustration of the fertility of matter-in-organization. Such a view would accord well with the Hebrew understanding of the psychosomatic unity of man. It would also accord with the Christian hope of resurrection, the reconstruction of that pattern, dissolved by death, in a new environment of God's choosing.[5]

* * *

---

[4] Quoted in K. Ward, *The Battle for the Soul* (London: Hodder & Stoughton, 1985), 34.

[5] The preceding section is from *Science and Creation* by John Polkinghorne © 1988 by John Polkinghorne, revised edition published by Templeton Press, 2006, 86–7.

Those who take the unified view of a material/mental world that I am advocating have sometimes felt driven to some form of panpsychism, the endowment of elementary matter with a residual prehensive power which in aggregation will lead to consciousness. David Griffin writes that 'the difference between the proton and the psyche is one of degree, not of kind (in an ontological sense). One who holds otherwise is a dualist, however odious such a description may be.'[6] Griffin is a process theologian, and the thought of A. N. Whitehead does indeed seem to have a panpsychic character. Events are basic to his metaphysics, and each event is held to have a quasi-subjective phase (prehension), followed by an objectification (concrescence), a sort of wedding of the material and mental in the marriage-bed of occurrence. This seems to me to be an unhappily literal way of seeking a synthesis. It is not the case in quantum theory that every particle has a little bit of undulation in it, which, when added together, gives a wave. The mixture is more subtle. The number operator (which counts the particles) and the phase operator (which specifies the wave) are what are called conjugate variables. They do not commute and so are incapable of simultaneous quantifiability.[7] A wave-like state is, therefore, one with an indefinite number of particles in it.

That insight cannot be applied by direct analogy to the case we are considering. The mental is certainly not characterized by association with an indefinite quantity of the material. If there is any connection whatever between the ideas of quantum physics and those of a complementary metaphysics, it might rather lie in the mental being associated with an indefinite degree of organization of the material, a sort of openness of pattern.[8]

* * *

Even if there is something in all this, it cannot be the whole story. The argument thus far has presented mind as the complementary pole of matter. While the discussion has been very far from thinking of mind as an epiphenomenon of matter, it nevertheless has inextricably linked the two, just as wave and particle are not to be divorced from each other in

---

[6] D. R. Griffin (ed.), *Physics and the Ultimate Significance of Time* (Albany, NY: State University of New York Press, 1986), 14.

[7] Polkinghorne, *The Quantum World*, 28.

[8] The preceding section is from *Science and Creation* by John Polkinghorne © 1988 by John Polkinghorne, revised edition published by Templeton Press, 2006, 88.

conventional quantum theory.[9] Yet we have good reason for supposing that there are inhabitants of the mental world that are not anchored in the material. The first candidates I would consider are the truths of mathematics. It is difficult to believe that they come into being with the action of the human mind that first thinks them. Rather, their nature seems to be that of ever-existing realities that are discovered, but not constructed, by the explorations of the human mind.[10]

---

[9] In the determinate version of quantum theory, using hidden variables, just such a divorce is made (see D. Bohm, *Wholeness and the Implicate Order* (London: Routledge & Kegan Paul, 1980), ch. 4; Polkinghorne, *The Quantum World*, 56–7). In its dualism of wave and particle, Bohm's theory is a sort of quantum Cartesianism.

[10] The preceding section is from *Science and Creation* by John Polkinghorne © 1988 by John Polkinghorne, revised edition published by Templeton Press, 2006, 90.

5

A brief history of science and religion

We form in a variety of ways the concepts which are the tools of our thought. Logical analysis is only a part of the process, for we also make use of the imaginative resource of story. Many people's attitude to science and religion is powerfully affected by two narratives, both of which appear to carry the message of a truth-seeking science confronting an obscurantist and conservative religion. These are the modern myths of Galileo and Charles Darwin, both seen as being in conflict with the Church. In the form in which these stories are deposited in contemporary minds, they are presented as simple accounts of the battle of light with darkness, an impression sedulously fostered by their re-presentation in the media from time to time. In actual fact, the truth is altogether more complex and correspondingly more interesting.

Galileo Galilei

Born in 1564, Galileo Galilei is unquestionably one of the great figures in the history of science. He repudiated mere appeal to the authority of Aristotle and in its place pioneered the investigative technique of combining mathematical argument with an appeal to observation and experiment. His brilliant use of the newly discovered telescope as a means for searching the heavens (resulting in the discovery of mountains on the Moon, spots on the Sun, satellites encircling Jupiter, and the phases of Venus) reinforced his belief in the Copernican system. By 1616, this had got him into trouble with the Vatican authorities, who believed that the Ptolemaic system, with its fixed Earth, was endorsed by the Bible. Some kind of accommodation was worked out between Galileo and his chief critic, Cardinal Bellarmine. The exact terms of this agreement later became a matter of dispute and there is a continuing scholarly debate on the question. The point at issue is whether Galileo was simply told not to espouse or defend the Copernican principle or whether he was also forbidden to teach it in any way whatsoever. Whatever the rights of the matter, intellectual freedom was clearly curtailed by the exercise of ecclesiastical authority.

In 1632, Galileo published his *Dialogue Concerning the Two Chief World Systems*. Cast in the apparent form of an even-handed discussion of the pros and cons of the ideas of Ptolemy and Copernicus, its actual presentation of the case for Copernicanism was so overwhelming that it was clearly a tract in that system's defence. Moreover, Simplicio, the defender of Ptolemy, was not only weak in argument and something of a buffoon, but he also stated, almost word for word, points of view which had been propounded by the current pope, Urban VIII. It is scarcely surprising that the authorities were upset and they responded by summoning Galileo to appear before them. He was sentenced by the Inquisition to life imprisonment, immediately commuted by the Pope to continuing house arrest. At no stage was Galileo subjected to torture.

No one can claim that this is an edifying story or that the church authorities displayed wisdom or intellectual integrity in their implacable opposition to Galileo's Copernican ideas. (The Roman Catholic ban on Copernicanism was rescinded in 1820 but Galileo's condemnation was only recently abrogated formally.) Yet the issues were complex and the illumination afforded by hindsight should not result in our painting the scene in stark black and white. There were scientific difficulties in the case presented by Galileo. One was the absence of stellar parallax – the shift in the apparent position of the stars expected to result from their being viewed from different perspectives if the Earth were moving round an orbit in the course of the year. (We now know that this was not observable with seventeenth-century resources because the stars are so very distant from us.) Galileo placed great emphasis on the claimed confirmatory value of his explanation of the tides. We now know that he was completely in error about this matter. He even ridiculed Kepler when the latter suggested that the Moon might have some relevance for tidal phenomena!

Throughout the controversy, and until his death, Galileo remained a religious man. Many of his discussions with his opponents had focused on the right way in which to read the Bible. Galileo genuinely valued its spiritual authority, but the fact that it was written in language intended to be understood by common people meant, in his opinion, that it was illegitimate to try to read advanced physical theory out of its pages. If there was an apparent conflict between the surface meaning of words of Scripture and the results of science, Galileo believed that this should encourage us to seek a deeper understanding of the relevant biblical passage – a view for which he could appeal to the support of St Augustine, no less.

Cardinal Bellarmine had urged upon Galileo the view that mathematical theories, like that of Copernicus, were just means of 'saving the appearances,' that is to say that they were calculational devices and not necessarily to be taken seriously as literal descriptions. Here we have an engagement with one of the fundamental questions in the philosophy of science, to which we shall subsequently return. Are scientific theories just convenient 'manners of speaking,' or do they describe the physical world as it actually is?

Finally, there were the personal aspects of the controversy: Urban VIII's wounded pride, Galileo's brilliant but polemically sharp use of the Italian language, the ambitions of Galileo's opponents among the Jesuit astronomers (to this day effective participants in the scientific community). These varied considerations do not mean that the Roman Catholic authorities did not make a bad mistake. Of course they did, but in complex and cloudy circumstances. The Galileo affair by no means indicates that there is an inevitable incompatibility between science and religion. One unwise incident does not imply a continuing conflict.

Charles Darwin

But do we not see the same thing happening all over again following the publication in 1859 of Charles Darwin's *Origin of Species*? Once more popular myth presents a picture of light confronting darkness. The image of Galileo before the Inquisition is succeeded by the image of Thomas Huxley vanquishing Bishop Samuel Wilberforce in their debate at the Oxford meeting of the British Association for the Advancement of Science in 1860. The story goes that the bishop was unwise enough to enquire of Huxley whether he was descended from an ape via his grandfather or his grandmother. Such a tasteless tactic brought the stern rebuke that Huxley would rather have an ape for an ancestor than a bishop who was unwilling to face the truth.

There is, in fact, some doubt about what actually happened on this occasion. Huxley's own version was put on paper thirty years after the event and the contemporary accounts are by no means unanimous in recounting a famous victory by the scientist. Be that as it may, once again the full story is more complex and confused than myth allows.

At the scientific level, there were contemporary biological critics of the idea of evolution by natural selection, like Sir Richard Owen, the greatest anatomist of the day, who pointed to difficulties in Darwin's thesis. Indeed Wilberforce himself, who was genuinely interested in scientific matters, wrote a review of the *Origin* which Darwin acknowledged as making some telling points in relation to the problems faced by the theory. The great British physicists of the nineteenth century, such as Faraday, Maxwell and Stokes, were silent in public but privately had doubts about the unaided adequacy of natural selection to explain the development of life in the timescale available. Lord Kelvin broke that silence when he claimed that the rate of the Earth's cooling and the length of the era during which the Sun could have been shining restricted the time available to a period much shorter than that required by Darwin's theory. While Kelvin's calculations were correct in terms of the known physics of his day, he was unaware of the processes of radioactivity (which has had a significant warming effect upon the Earth) and nuclear fusion (which has powered the Sun for the five billion years of its shining).

If the scientific scene was confused, so was the religious. At the very same meeting of the British Association which had seen the debate between Wilberforce and Huxley, Frederick Temple (later to be Archbishop of Canterbury) preached a sermon welcoming the insights of evolution. There was by no means uniform opposition to Darwin's ideas from within the Church. Charles Kingsley took a robust view of accepting scientific truth and insight, seeing natural selection as relating to the 'how' of God's creative action and interpreting evolution as replacing the notion of the Creator's instant act by the subtler and more satisfying idea of a creation brought into being and able then to 'make itself.' One of Darwin's friends and regular correspondents, Asa Gray the Harvard botanist, did much to make evolution a respectable idea among thinking people in North America, while remaining a deeply religious man.

Once again, there were personal factors at work, influencing the behaviour of the participants. Wilberforce may have wanted to stand on episcopal dignity, but Thomas Huxley was also strongly motivated by nonintellectual considerations, such as the desire to reduce the traditional influence of the clergy and to establish the authority of the newly emerging class of professional scientists. Charles Darwin's own loss of the Christian belief he had held as a young man is thought to have been at least as much influenced by sadness at the harrowing death of his daughter Annie at the age of ten, as by his scientific discoveries. In assessing Darwin's later cautious utterances on religion one must remember his sensitive wish not to offend his wife Emma, who was a person of religious faith, but he never became an out-and-out atheist. Even Huxley did not go as far as explicit atheism, coining the word 'agnostic' to describe those who, like himself, felt the question of God's existence to be beyond settlement.

Conclusion

Rightly read, the Galileo and Darwin incidents are instructive and focused examples of how religion and science can interact. In each case, certain beliefs previously held by all people (the centrality of the Earth, the immutability of species) proved to be in need of radical revision. Because theological thinking had incorporated these common notions into its background assumptions, it too had to make changes in how it understood its insights to relate to other forms of knowledge. This was not a comfortable process for theology. In each case, there was initial resistance, but this was by no means total. At the same time, the scientific case was not itself immediately crystal clear and arguments persisted for a while, since radical revision is no more easy for scientists than it is for theologians. Yet in the end the dust settled for both subjects. Theology discovered that the dignity of humankind depended neither upon its inhabiting the centre of the universe nor upon *Homo sapiens* being a separately and instantaneously created species.

Some scholars have even suggested that, far from science and religion being at enmity with each other, it was the Judaeo-Christian-Islamic concept of the world as creation that enabled science to flower in seventeenth-century Europe, rather than in ancient Greece or medieval China, despite the great intellectual achievements of these latter two civilizations. The doctrine of creation implies that:

- the world is orderly, since God is rational;
- no prior constraints are imposed on the Creator's choice of creation's pattern, so that one has to look (observe and experiment) to see what the divine will has selected;
- because creation is not itself sacred, it can be investigated without impiety;
- because the world is God's creation, it is a worthy object of study.

Certainly the search for order through experimental investigation is fundamental to the practice of science and not all religious cultures would provide encouragement to this task.[1]

---

[1] The preceding section is from *Science and Theology* by John Polkinghorne © 1998 by Fortress Press, published in the USA by Fortress Press, reprinted by permission of Augsburg Fortress Publishers, 4–9.

6

Science and religion as cousins

People sometimes think that it is odd, or even disingenuous, for a person to be both a physicist and a priest. It induces in them the same sort of quizzical surprise that would greet the claim to be a vegetarian butcher. Yet to someone like myself who is both a scientist and a Christian, it seems to be a natural and harmonious combination. The basic reason is simply that science and theology are both concerned with the search for truth. In consequence, they complement each other rather than contrast one another. Of course, the two disciplines focus on different dimensions of truth, but they share a common conviction that there is truth to be sought. Although in both kinds of enquiry this truth will never be grasped totally and exhaustively, it can be approximated to in an intellectually satisfying manner that deserves the adjective 'verisimilitudinous,' even if it does not qualify to be described in an absolute sense as 'complete.'

Certain philosophical critiques notwithstanding, the pursuit of truthful knowledge is a widely accepted goal in the scientific community. Scientists believe that they can gain an understanding of the physical world that will prove to be reliable and persuasively insightful within the defined limits of a well-winnowed domain. The idea that nuclear matter is composed of quarks and gluons is unlikely to be the very last word in fundamental physics – maybe the speculations of the string theorists will prove to be correct, and the quarks, currently treated as basic constituents, will themselves turn out eventually to be manifestations of the properties of very much smaller loops vibrating in an extended multidimensional spacetime – but quark theory is surely a reliable picture of the behaviour of matter encountered on a certain scale of detailed structure, and it provides us with a verisimilitudinous account at that level.

Theologians entertain similar aspirations. While the infinite reality of God will always elude being totally confined within the finite limits of human reason, the theologians believe that the divine nature has been revealed to us in manners accessible to human understanding, so that these self-manifestations of deity provide a reliable guide to the Creator's relationship with creatures and to God's intentions for ultimate human fulfilment. For the Christian, this divine self-revelation centres on the history of Israel and the life, death and resurrection of Jesus Christ, foundational events that are the basis for continuing reflection and exploration within the Church, an activity that the community of the faithful believes to be undertaken under the guidance of the Holy Spirit. Revelation is not a matter of unchallengeable propositions mysteriously conveyed for the unquestioning acceptance of believers, but it is the record of unique and uniquely significant events of divine disclosure that form an indispensable part of the rational motivation for religious belief.[1]

* * *

Thus, I see there to be a cousinly relationship between the ways in which theology and science each pursue truth within the proper domains of their interpreted experience. Critical realism is a concept applicable to both, not because there is some kind of entailment from method in one to method in the other – for the differences in their subject material would preclude so simple a connection – but because the idea is deep enough to encompass the character of both these forms of the human search for truthful understanding.

This is a theme that I have often discussed in my writing. Pursuing it requires the analysis of actual examples, rather than relying on an attempted appeal to grand general principles. In my Terry Lectures, I sought to set out five points of analogy between two seminal developments, one in physics and one in Christian theology: the exploration of quantum insight and the exploration of Christological insight.[2] In making this comparison, I discerned five points of cousinly relationship between these two great human struggles with the surprising and counterintuitive character of our encounter with reality. In outline, these five points are:

1 Moments of enforced medical revision

The crisis in physics that led eventually to quantum theory began with great perplexity about the nature of light. The nineteenth century had shown quite decisively that light possessed wave-like properties. However,

---

[1] The preceding section is from *Quantum Physics and Theology* by John Polkinghorne © 2007 by Yale University, published in the USA by Yale University Press, reprinted by permission of Yale University Press, 1–3.

[2] J. C. Polkinghorne, *Belief in God in an Age of Science* (Yale University Press, 1998), ch. 2.

at the start of the twentieth century, phenomena were discovered that could only be understood on the basis of accepting the revolutionary ideas of Max Planck and Albert Einstein that treated light as sometimes behaving in a particle-like way, as if it were composed of discrete packets of energy. Yet the notion of a wave/particle duality appeared to be absolutely nonsensical. After all, a wave is spread out and oscillating, while a particle is concentrated and bullet-like. How could anything manifest such contradictory properties? Nevertheless, wave/particle duality was empirically endorsed as a fact of experience, and so some radical rethinking was evidently called for. After much intellectual struggle this eventually led to modern quantum theory.[3]

In the New Testament, the writers knew that when they referred to Jesus they were speaking about someone who had lived a human life in Palestine within living memory. Yet they also found that when they spoke about their experiences of the risen Christ, they were driven to use divine-sounding language about him. For example, Jesus is repeatedly given the title 'Lord,' despite the fact that monotheistic Jews associated this title particularly with the one true God of Israel, using it as a substitute for the unutterable divine name in the reading of scripture. Paul can even take verses from the Hebrew Bible that clearly refer to Israel's God and apply them to Jesus (for example, compare Philippians 2.10–11 with Isaiah 45.23, and 1 Corinthians 8.6 with Deuteronomy 6.4). How could this possibly make sense? After all, Jesus was crucified and Jews saw this form of execution as being a sign of divine rejection, since Deuteronomy (21.23) proclaims a curse on anyone hung on a tree. Experience and understanding seemed as much at odds here as they did in the case of the physicists' thinking about light.

2 A period of unresolved confusion

From 1900 to 1925, the physicists had to live with the paradox of wave/particle duality unresolved. Various techniques for making the best of a baffling situation were invented, by Niels Bohr and others, but these expedients were no more than patches clapped onto the broken edifice of Newtonian physics, rather than amounting to the construction of a grand new quantum building. It was intellectually all very messy, and many physicists at the time simply averted their eyes and got on with the less

---

[3] For an introduction to quantum theory, see J. C. Polkinghorne, *Quantum Theory: A Very Short Introduction* (Oxford University Press, 2002).

troubling task of tackling detailed questions that were free from such fundamental difficulties. Problem-solving in normal science is often a more comfortable pursuit than wrestling with perplexities in revolutionary science.

In the New Testament, the tension between human and divine language used about Jesus is simply there, without any systematic theological attempt being made to resolve the matter. It seems that those early generations of Christians were so overwhelmed by the new thing that they believed that God had done in Christ, that its authenticity and power were of themselves sufficient to sustain them without forcing them to attempt an overarching theoretical account. Yet, the position taken by those New Testament writers was clearly intellectually unstable, and the issue could not be ignored indefinitely.

3 New synthesis and understanding

In the case of physics, new insight came with startling suddenness through the theoretical discoveries of Werner Heisenberg and Erwin Schrodinger, made in those amazing years, 1925–6. An internally consistent theory was brought to birth, which required the adoption of novel and unanticipated ways of thought. Paul Dirac emphasised that the formal basis of quantum theory lay in what he called the superposition principle. This asserts that there are quantum states that are formed by adding together, in a mathematically well-defined way, physical possibilities that Newtonian physics and commonsense would hold to be absolutely incapable of mixing with each other. For example, an electron can be in a state that is a mixture of 'here' and 'there,' a combination that reflects the fuzzy unpicturability of the quantum world and which also leads to a probabilistic interpretation, since a 50–50 mixture of these possibilities is found to imply that, if a number of measurements of position are actually made on electrons in this state, half the time the electron will be found 'here' and half the time 'there.' This counterintuitive principle just had to be accepted as an article of quantum faith. Richard Feynman introduced his lectures on quantum mechanics by talking about the two-slits experiment (a striking example of counterintuitive quantum ambidexterity), concerning which he wrote,

> Because atomic behaviour is so unlike ordinary experience, it is very difficult to get used to, and it appears peculiar and mysterious to everyone . . . we shall tackle immediately the basic element of the mysterious behaviour in its most strange form. We choose to examine a phenomenon which is impossible, *absolutely* impossible, to explain in any classical way, and which has

in it the heart of quantum mechanics. In reality it contains the *only* mystery. We cannot make the mystery go away by 'explaining' how it works. We will just *tell* you how it works.[4]

The quest for a deeper understanding of the fundamental phenomena recorded in the New Testament, eventually led the Church to a trinitarian understanding of the nature of God (Councils of Nicaea, 325, and Constantinople, 381) and to an incarnational understanding of two natures, human and divine, present in the one person of Christ (Chalcedon, 451). These were important Christian clarifications, but one cannot claim that theology, wrestling with its profound problem of understanding the divine, has been as successful as science has been in attaining its understanding of the physical world. The latter is at our disposal to interrogate and put to the experimental test, but the encounter with God takes place on different terms, involving awe and worship and obedience. There is an important qualifying theological insight, called apophatic theology, stressing the otherness of God and the necessary human limitation in being able to speak adequately of the mystery of the divine nature. There are bounds to the possibilities of theological explanation. The Fathers of the Church, who at the Councils had formulated fundamental Christian insights, would, I believe, have been quite content to echo Feynman's words, 'We will just *tell* you how it works.'

4 Continued wrestling with unsolved problems

Even in science, total success is often elusive. Quantum theory has been brilliantly effective in enabling us to do the sums, and their answers have proved to be in extremely impressive agreement with experimental results. However, some significant interpretative issues still remain matters of uncertainty and dispute. Chief among these is the so-called measurement problem. How does it come about that a particular result is obtained on a particular occasion of measurement, so that the electron is found to be 'here' this time, rather than 'there'? It is embarrassing for a physicist to have to admit that currently there is no wholly satisfactory or universally accepted answer to that entirely reasonable question. Quantum physics has had to be content for eighty years to live with the uncomfortable fact that not all its problems have yielded to solution. There are still matters that we do not fully understand.

---

[4] R. Feynman, *The Feynman Lectures on Physics*, vol. 3 (Redwood City, Calif.: Addison-Wesley, 1965), 7.

Theology also has had to be content with a partial degree of understanding. Trinitarian terminology, for example in its attempt to discriminate the divine Persons in terms of a distinction between begetting and procession, can sometimes seem to be involved in trying to speak what is ineffable. The definition of Chalcedon, asserting that in Christ there are two natures 'without confusion, without change, without division, without separation,' is more a statement of criteria to be satisfied if Christological discourse is to prove adequate to the experience preserved in scripture and continued within the Church's tradition, than the articulation of a fully developed Christological theory. Chalcedon maps out the enclosure within which it believes that orthodox Christian thinking should be contained, but it does not formulate the precise form that thinking has to take. In fact, further Christological argument, both within the Chalcedonian bounds and outside them, has continued down the centuries since 451.

5 Deeper implications

A persuasive argument for a critical realist position lies in its offering an explanation of how further successful explanations can arise from a theory, often concerning phenomena not explicitly considered, or even known, when the original ideas were formulated. Such persistent fruitfulness encourages the belief that one is indeed 'on to something,' and that a verisimilitudinous account has been attained. In the case of quantum theory, a number of successes of this kind have come to light, including explaining the stability of atoms (their remaining unmodified by the numerous low-energy collisions to which they are subjected), and the very detailed calculations of their spectral properties that have proved to be in impressive agreement with experimental measurements. Strikingly novel, and eventually experimentally verified, predictions have also been made. One of the most outstanding of these is the so-called EPR effect, a counterintuitive togetherness-in-separation that implies that two quantum entities that have interacted with each other remain mutually entangled, however far they may subsequently separate in space. Effectively, they remain a single system, for acting on the one 'here' will produce an immediate effect on its distant partner.

Incarnational belief has offered theology some analogous degree of new insight. For example, Jürgen Moltmann has made powerful use of the concept of divine participation in creaturely suffering through the cross of Christ. He emphasizes that the Christian God is the crucified God,[5] the

---

[5] J. Moltmann, *The Crucified God* (London: SCM Press, 1974).

One who is not just a compassionate spectator of the suffering of creatures but a fellow-sharer in the travail of creation. The concept of a suffering God affords theology some help as it wrestles with its most difficult problem, the evil and suffering present in the world.[6]

---

[6] The preceding section is from *Quantum Physics and Theology* by John Polkinghorne © 2007 by Yale University, published in the USA by Yale University Press, reprinted by permission of Yale University Press, 15–22.

7

The work of love

A prayer in the current Daily Office of the Church of England begins, 'Eternal God and Father, you create us by your power and redeem us by your love. . . .' Although it is a much-loved prayer, its theology is open to question. We could as well speak of the God who creates us by divine love and redeems us by divine power.

Any dichotomy between creation and redemption carries with it theological dangers, and these risks are enhanced when there is an imposed correlation with different divine attributes. The act of creation, of bringing a world into being and maintaining it in being, is clearly an act of great power to which the puny powers of creatures bear no comparison. In theological discourse, only God can provide the answer to the great question, 'Why is there something rather than nothing?' But that question is not solely answered by citing the divine power of radical creation out of nothing. If we may use such language, it is also necessary to consider, so to speak, what are God's motives that lie behind this great act. Pursuing that point surely involves appeal to the divine love that has willed the existence of the truly other so that, through creation, this love is also bestowed outside the perichoretic exchange between the Persons of the Holy Trinity. Creation exists because God gives to it a life and a value of its own.

The Christian hope of redemption from bondage to sin and futility, both for individual human beings and for the whole created order, is certainly founded on trust in the unfailing faithfulness of the God of love (*chesed*). Jesus made just such a point in his dispute with the Sadducees about whether there is a destiny beyond death. He affirmed that the God of Abraham, Isaac, and Jacob cares everlastingly for the patriarchs and so is 'not God of the dead, but of the living' (Mark 12.18–27). Yet the context for redemption, both human and cosmic, is the new creation. The latter is not a second creative act *ex nihilo*, for it proceeds *ex vetere* as the resurrected transformation of the old creation.[1] Differing in this respect from

---

[1] J. C. Polkinghorne, *Science and Christian Belief/The Faith of a Physicist* (SPCK/Princeton University Press, 1994), ch. 9.

the initial creation, the new creation nevertheless exhibits the working of great divine power of a transcendent kind. That power has, in fact, already been manifested in the resurrection of Jesus Christ (cf. Romans 1.4).

Love without power would correspond to a God who was a compassionate but impotent spectator of the history of the world. Power without love would correspond to a God who was the Cosmic Tyrant, holding the whole of history in an unrelenting grasp. Neither would be the God and Father of our Lord Jesus Christ, for the Christian God can neither be the Creator of a divine puppet theater nor a deistic Bystander, watching the play of history unfold without any influence upon its course. Divine power and divine love must both be given their due importance. The title of this paper, juxtaposing kenosis and action, strives for the necessary balance and so poses the problem of how it is to be attained. The solution of that problem is not an easy task to achieve. All theological thinking is a precarious balancing act, seeking recourse to the coincidence of opposites in an attempt to use finite human language to discourse about the infinite reality of God. Every assertion seems to stand in need of the qualification of a counter-assertion. The warnings of apophatic theology need to be heeded, but not to the extent of a total paralysis of thought.

The need to do justice both to divine kenotic love and to divine providential power is clearly part of this theological tension. Emphasis on divine love seems to lie behind Process Theology's picture of a God who, in A. N. Whitehead's moving phrase, is a 'fellow sufferer who understands,' and who acts only through the power of persuasion. It is a noble concept, but it is open to question whether deity has not been so evacuated of power that hope in God as the ground of ultimate fulfillment has been subverted. The issue is whether the presentation of the divine vision of fulfillment will be sufficient in itself to bring about its own realization, or whether creation also stands in need of the action of divine grace for this to be achieved. The matter can be put in the bluntest terms by asking whether Whitehead's God could be the One who raised Jesus from the dead.

Emphasis on divine power seems to lie behind Classical Theology's picture of a God who, through primary causality, is in total control and whose invulnerability is such that there is no reciprocal effect of creatures upon the divine nature, of the kind that a truly loving relationship would seem to imply. The scheme, as articulated by its principal exponents such as Aquinas, is intellectually impressive, but it is open to question whether its picture of the divine nature is not so remote and insulated from creation as to put in question the fundamental Christian conviction that 'God is love' (1 John 4.8). There are also unresolved difficulties about the coherence of supposing that divine primary causality and creaturely secondary causality are both simultaneously at work in the world.

A great deal of creative theological thinking in the second half of the twentieth century has been concerned with a re-examination of these issues. A number of factors have encouraged this. Some of them are theological in character and we shall be content simply to indicate their general nature, since they are major concerns of other contributors to this symposium. Others arise, at least partly, from scientific insights, and we shall give them greater attention.

1 Incarnational theology

The classic kenotic text in the New Testament is Philippians 2.1–11, referring to Christ who 'emptied (*ekenosen*) himself, taking the form of a servant.' The self-limitation of the divine Word, in taking flesh and becoming a finite human being, is a concept that has sometimes powerfully influenced christological thinking. That was so among Lutherans in the seventeenth and early nineteenth centuries, and again among certain British theologians at the end of the nineteenth century. In the twentieth century, the application of kenotic ideas has been extended beyond a strictly christological focus to include other aspects of God's relationship with creation. Jürgen Moltmann has powerfully laid before us the concept of the Crucified God, revealed in the Trinitarian event of the cross of Jesus Christ.[2] W. H. Vanstone has made kenotic concepts central to his exploration of the precariousness, costliness, and gift of value involved in loving acts of creativity.[3] We shall return to Incarnational insight later.

2 Theodicy

The classic location for confrontation between the claims of divine love and divine power has always been the perplexities of theodicy. Divine love would seem to imply a Creator whose benevolent wish would be for the total goodness of creation. Divine power would seem to imply a Creator who can accomplish fully the divine purposes. Whence then have come the many physical ills of disease and disaster? No doubt the existence of moral beings with the freedom to choose how they act is a great good. The misuse of this gift can be seen as the origin of moral evil. However,

---

[2] J. Moltmann, *The Crucified God* (London: SCM Press, 1974).
[3] W. H. Vanstone, *Love's Endeavour, Love's Expense* (London: Darton, Longman and Todd, 1977).

we can no longer believe with Augustine that this abuse has also had consequences of cosmic scope for all creation, leading to the corruption of what had previously possessed paradisal perfection, so that it was the source of the physical evil in the world. Much of the problem of evil, both moral and physical, lies in its scale. In the century of the Holocaust and Hiroshima, we are only too conscious of these issues. Moltmann's work is explicitly a contribution to theology after Auschwitz, and Vanstone is concerned with the necessary precariousness of the creative process of the world. Again, these are issues to which we shall return.

3 Continuous creation

From the later eighteenth century onwards, there has been a progressive scientific unfolding of the historical character of the physical world. In terrestrial terms, it was first realized that the landscape of Earth had been subject to eons of gradual change. Some mountains had been eroded and others thrust up by subterranean process to take their place. Eventually it was discovered that the great land masses themselves had been subject to continental drift. Our impression of stasis arises simply from the shortness of human life, and of recorded history, in relation to the time scales of geological change. From a longer temporal perspective, the landscape is as much in flux as the cloudscape.

In the nineteenth century, the long-assumed fixity of species came under critical reassessment, culminating in 1859 with the publication of the evolutionary ideas of Charles Darwin and A. R. Wallace. The biosphere turned out to be no more static than the geosphere.

In the first decades of the twentieth century, against the ideological inclinations of many scientists (including Einstein), it was realized that the universe itself has also had a changing history. According to big bang cosmology, there was a time when there were no stars and galaxies. We live on a second-generation planet, encircling a second-generation star, which have both condensed from gas clouds and the debris of first-generation supernovae explosions.

These scientific revaluations have had their impact upon theology. The effects have been by no means without their fruitfulness. Contrary to the popular legend, still sedulously cultivated by polemicists and some sectors of the media, that Darwin met total obscurantist opposition from religious thinkers, there were Christians who, from the first, welcomed his insights and made positive theological use of them. In England, these included Charles Kingsley, Frederick Temple, and Aubrey Moore, and in the United States Darwin's Harvard friend, Asa Gray. A common theme runs through their responses. An evolutionary world is theologically to be understood as a creation allowed by its Creator 'to make itself.' The play of life is not the performance of a pre-determined script, but a self-improvisatory performance by the actors themselves. Although kenotic language was not explicitly used, this is a manifestly kenotic conception. God shares the unfolding course of creation with creatures, who have their divinely allowed, but not divinely dictated, roles to play in its fruitful becoming.

This understanding is in striking contrast with the accounts of creation and divine action offered by Classical Theology. From Augustine onwards, and most powerfully in the writings of Thomas Aquinas, it sought to preserve the uniqueness of divine action, and the primacy of divine power, by speaking of God's primary causality, exercised in and under the secondary causalities of creatures. Classical Theology greatly emphasized the transcendence of God and a characteristic concept was creation *ex nihilo*, the calling into being of the new at the behest of the divine creative fiat.

Of course, the distinction between a transcendent Creator, possessed of aseity, and a creation that is perpetually dependent on the will of God for its preservation from the abyss of nothingness, is a very important component in Christian theological thinking. We are right to be extremely cautious about any ideas that might put this in question. Only a God who is distinct from creation can be that creation's ground of hope beyond its eventual natural decay. However, it by no means follows that in saying this we have said all that needs to be said about the doctrine of creation. We should remember that even Augustine, reflecting on Genesis 1, vv. 11, 20, and 24, in which the earth and the waters bring forth life, believed that God had created the seeds of life from which creatures eventually developed, although, of course, he had no concept of the transformation of one species into another.

The scientific recognition of the evolutionary character of the universe has encouraged theological recognition of the immanent presence of God to creation and of the need to complement the concept of *creatio ex nihilo* by a concept of *creatio continua*. Continuous creation has been an important theme in the writings of the scientist-theologians.[4] It has a number of important theological implications.

First, in conjunction with the understanding of evolutionary process as corresponding to creation being allowed to make itself, it is clearly

---

[4] I. G. Barbour, *Issues in Science and Religion* (London: SCM Press, 1966), ch. 12; A. R. Peacocke, *Creation and the World of Science* (Oxford: Oxford University Press, 1979), chs 2 and 3; J. C. Polkinghorne, *Science and Creation* (SPCK, 1988), ch. 4.

kenotic in its character. Its unfolding process is to be understood as being flexible and open to creaturely causality. Philip Hefner, whose theology is strongly influenced by evolutionary ideas, likes to speak of human beings as 'created co-creators.'[5]

Second, this kenotic sharing of power has important implications for theodicy. No longer can God be held to be totally and directly responsible for all that happens. An evolutionary world is inevitably one in which there are raggednesses and blind alleys. Death is the necessary cost of new life; environmental change can lead to extinctions; genetic mutations sometimes produce new forms of life, oftentimes malignancies. There is an unavoidable cost attached to a world allowed to make itself. Creatures will behave in accordance with their natures: lions will kill their prey; earthquakes will happen; volcanoes will erupt and rivers flood. I have called this insight 'the free-process defense'[6] in relation to physical evil, in analogy with the familiar free-will defense in relation to moral evil. These defenses do not by any means solve all the problems of theodicy, but they temper them somewhat by removing a suspicion of divine incompetence or indifference. From this point of view, the classic confrontation between the claims of divine love and the claims of divine power is resolved by maintaining God's total benevolence but qualifying, in a kenotic way, the operation of God's power. Of course, this is a self-qualification, exercised within the divine nature and in accordance with that nature itself. It is quite different from Process Theology's conception of an external metaphysical constraint upon the power of deity, for in this case it is held that nothing imposes conditions on God from the outside. The classical theologians were right in that respect, but they had not taken adequately into account the interior 'constraints' of the self-consistency of the divine nature. Perhaps their strong emphasis on divine Unity made such a consideration inaccessible to them.

Third, if the concept of continuous creation is really to mean what it says, and to consist of more than just a pious gloss on a wholly natural process, then God's providential guiding power must surely also be part of the unfolding of evolutionary history. The kenotic Creator may not overrule creatures, but the continuous Creator must interact with creation. Thus, kenotic *creation* and divine *action* are opposite sides of the same theological coin. This is not to deny that the natural process in itself is also an expression of the Creator's will, for it corresponds to that general

---

[5] P. J. Hefner, *The Human Factor* (Minneapolis: Fortress Press, 1993), *passim*.
[6] J. C. Polkinghorne, *Science and Providence* (SPCK, 1989), 66–7.

providence that is manifested in the divine ordinance of natural law. However, a notion of continuous creation may be expected to go beyond a deistic upholding of the universe in being, for so strong a concept seems inadequately realized in terms of the God of natural theology alone, who is simply the ground of cosmic order. Putting it another way, if, as is often said, evolutionary process is generated by the interplay between 'chance' (that is, historical contingency) and 'necessity' (that is, lawful regularity), its Creator must be present in the contingency as well as in the regularity.

This conclusion is reinforced by considering why personal language is used in the Judeo-Christian tradition as the least misleading way of referring to God. It is surely because the God addressed as Father is one who is expected to do particular things on particular occasions and not just to function as an unchanging effect (like the law of gravity). Clearly there is a problem here which we shall have to discuss further shortly. On the one hand, we have science's account of the regularity of the processes of nature. On the other hand, we have theology's claim to speak of a God who acts in history. Can the two be reconciled with each other? I believe so, but achieving this end will call for some flexibility from both science and theology in the assessments that they initially bring to their dialogue.

4 Causal nexus

If divine providential action is an integral part of continuous creation, so that the latter is understood in some sense as being evolution in which a degree of theistic guidance and influence is present, it is necessary to consider what we know about the causal nexus of the world, with a view to seeing if it could possibly accommodate such an idea. Science has some things to say about this, but from the point of view of Classical Theology, they would not be important.

The reason is that, for Thomist thinkers, divine primary causality is considered to be of a totally different kind from the creaturely causal powers that science investigates, for it is held to be exercised in and under those manifold secondary causalities. No explanation is offered of how this happens; it is simply said to be the case. Any attempt to exhibit the 'causal joint' by which the double agency of divine and creaturely cau-salities are related to each other is held to be impossible, or even impious. Three assertions are important consequences of this point of view.

First, the ineffability of the mode of action of primary causality has the effect of totally repudiating the possibility of any analogy between divine and human agencies. Second, God is party to every event not simply by allowing it to happen, but also by bringing it about through the exercise of the divine will. Nothing is outside God's control, an assertion that poses obvious difficulties for theodicy. The veiled and mysterious nature of primary causality can only be matched by the veiled and mysterious claim that in the end all will be seen to have been well. Third, primary causality is so divorced from the character of secondary causality that it may be believed to be active, whatever form the latter is discovered by science to take. Theology is made invulnerable to whatever may be currently understood about the process of the physical universe. However, some of us feel that the deep obscurity involved in the idea of a double agency carries with it the danger that the discussion might turn out to be no more than double talk.

What for its partisans is the strength of the notion of primary and secondary causalities is, for its critics, its greatest weakness. The strategy represents an extreme case of the 'two languages' way of understanding how science and theology relate to each other. Their discourses are treated as independent, so that they talk past each other at different levels. The two disciplines may be considered as presenting two different paradigms, or involving participation in two different language games. This is a point of view that the scientist-theologians,[7] together with many others working at the interface of science and theology, are rightly emphatic in rejecting. In its place they put the affirmation of the unity of truth and knowledge – a unity ultimately guaranteed by the oneness of the Creator – with its implication that there is an active intercourse across the boundary between the two disciplines. Creation is not so distanced from its Creator that the character of its process and history offers no clues at all to the nature of God's interaction with it.

Human and divine agency both clearly fall into the category of experience that is presently well beyond our capacity for full understanding. As persons, we should not deny our basic experiences of free choice and consequent moral responsibility. As Christians, we should not deny our intuition, and the testimony of our tradition, that God acts in the world. As rational thinkers convinced of the unity of knowledge, we should not forgo the attempt, however modest and tentative it must necessarily be, to see if metaphysical conjecture, arising out of modern science's understanding of physical process, might not afford us some small purchase on these problems.

---

[7] See J. C. Polkinghorne, *Scientists as Theologians* (SPCK, 1996), ch. 1.

Modern physics has revealed the existence of extensive *intrinsic* unpredictabilities, both in the microscopic realm of quantum physics and also in the macroscopic realm of chaotic dynamics. Such intrinsic epistemological deficiencies offer the chance, if we are bold enough to take it, of trying to turn them into ontological opportunities of openness, treating them as potential loci for the operation of additional causal principles, active in bringing about the future and going beyond and complementing the causal principle of the exchange of energy between constituents that has been the conventional description used by physics. The complete set of all causal principles, including human and divine agency, will then be what brings about the future state of the world.

Such a move is metaphysical in character; it can neither be affirmed nor denied on the basis of science alone. Since scientists are instinctive realists, such an attempt, which amounts essentially to aligning epistemology (unpredictability) and ontology (new causal principles) as closely as possible to each other, is a natural strategy for them to pursue. In the case of quantum theory, this has been the almost universal tactic, with Heisenberg's uncertainty principle (which originally related to the epistemological issue of what can be measured) being interpreted ontologically as a principle of indeterminacy, and not just a principle of ignorance. The alternative ideas of David Bohm, yielding a deterministic interpretation of quantum theory,[8] make it clear that this is a metaphysical choice and not a logically forced move. Some of us have sought to pursue a similar tactic in relation to chaotic dynamics.[9,10]

---

[8] D. Bohm and B. J. Hiley, *The Undivided Universe* (London: Routledge, 1993).

[9] I. Prigogine, *The End of Certainty* (New York: The Free Press, 1996); J. C. Polkinghorne, *Reason and Reality* (SPCK, 1991), ch. 3; *Belief in God in an Age of Science* (Yale University Press, 1998), ch. 3.

[10] The preceding section is from *The Work of Love* by John Polkinghorne © 2001 Wm. B. Eerdmans Publishing Co., published in the USA by Wm. B. Eerdmans Publishing Co., reprinted by permission of Wm. B. Eerdmans Publishing Co.

Part 2
GOD

8

The nature of theology

Scientists often use the word 'theological' in a pejorative sense, implying the absence of rigor and the presence of unmotivated assertion. Those who speak thus have a mental picture of theologians shutting their eyes and gritting their teeth in the effort to defend the indefensible, crying with Tertullian, 'I believe because it is absurd.' Paul Davies articulates a common view when he declares, 'The true believer must stand by his faith whatever the evidence against it.'[1] He also says, 'Religion is founded on dogma and received wisdom which purports to represent immutable truth.'[2]

There is sufficient half-truth in these statements to make them dangerously misleading. If there is a God, he is a hidden God. He does not make himself known unambiguously in acts of transparent significance, invariably preserving those who trust him from every misfortune and regularly restraining and punishing the acts of transgressors. Neither prayer nor blasphemy is a magical lever which can be used to act upon God to make him demonstrate his existence. He is not to be put to the test,[3] either by the demand for a particular outcome or by challenge to his authority. If man has been given independence so that he may freely choose his response to God, this elusive character seems necessary in One whose infinite presence, totally disclosed, would overwhelm our finite being.

Religious experience has a mysteriously open character. The believer is ill and prays. If he recovers, he thanks God for his healing; if he does not, he seeks to accept that also as the will of God. Either way he believes he has received wholeness, given by the sustaining grace of God, whose exact nature is to be found only within the experience itself. The unbeliever may exclaim in exasperation, 'Is God's head never on the block? Is it always "heads he wins, tails you lose"?' The brilliant mistranslation of the Authorized Version does not accurately render the Hebrew, but it expresses exactly an element of the religious man's experience when it has Job say,

---

[1] P. Davies, *God and the New Physics* (London: Dent, 1983), 6.
[2] Ibid., 220.
[3] Deuteronomy 6.16.

79

'Though he slay me, yet will I trust in him.'[4] To acknowledge this is to do justice to the character of what is involved in religious understanding, to recognize that it has its own nature which has to be respected.

Yet in the appropriate terms, religious conviction is still to be evaluated. It may be impossible to lay down beforehand exactly what those terms are (just as we cannot give a priori specification of when the use of inductive arguments is scientifically justified). Only in the event itself can its meaning be found. Yet it is not the case that *whatever* the evidence that event yields, the believer will be uncritical in his assessment of it. The psalms contain many protests to God.

Tradition certainly plays an important part in religion. So it does in science. We inherit the legacy of those who have preceded us, and it would be disastrous if every generation had to start from scratch. I remember visiting a laboratory in Eastern Europe where Maxwell's equations (the fundamental equations of electromagnetic theory) were engraved on tablets of stone in conscious imitation of the Ten Commandments. Dogma (the root meaning of the word is 'that which seems to be the case') and received wisdom are a necessary part of all human activity. The greater the role of personal judgment in a subject, the more we need the correctives afforded us by insights from the past. In that way we can best hope to allow for the tricks of intellectual perspective induced by the cultural conditioning of the present day.

Science is least vulnerable in this way, and that is why its achievements present the cumulative character of increases in knowledge. The more personal the subject, the greater the risk that we are prisoners in the cultural cage of contemporary attitudes. The men of the past may have known things which are necessary but which we have lost sight of. For example, St Augustine in the late fourth century understood the complex structure of the human psyche, with its internal polarities, in a way which has only recently been rediscovered through the insights of depth psychology.

Moreover, science deals in generalities, in principle accessible to all; more personal forms of knowledge are also concerned with the illumination given to and through a particular individual at a particular time. Hence the so-called scandal of particularity, the emphasis on the unique. Our musical understanding and experience would be greatly impoverished if we balked at the scandal of the particularity of J. S. Bach and refused his musical offering conveyed to us in the tradition. In the same way religion looks to the insights of the spiritual masters. In the sphere of the

---

[4] Job 13.15 (KJV).

personal, it is not inconceivable that the truest understanding of God is to be found in the possession of a wandering carpenter in a peripheral province of the Roman Empire, far away and long ago.

Yet respect for the wisdom of other ages does not imply an idolatrous servitude to it. The Christian creeds are summaries of the church's insights into her experience of God, but each generation has to make them its own to the extent that it can. Theology, like science, is corrigible. There is nothing immutable in its pronouncements. If they are found wanting after careful investigation, then they are to be abandoned. Theology has long understood the distinction between truth and verisimilitude; every image of God is an idol which eventually has to be broken in the search for Reality.

The view of the theological enterprise which I would wish to defend is summed up in a splendid phrase of St Anselm: *fides quaerens intellectum*, faith seeking understanding.[5]

* * *

The fact of the matter is that there are widespread claims to the experience of a religious dimension to reality; of encounter with the numinous presence of an Other; the recognition of unity with a reality transcending oneself; the perception of a purpose at work in the world that carries the assurance (all things to the contrary notwithstanding) that all shall be well; the acknowledgment of an ultimate significance to be found in the way the world is. Of course, one must also recognize great variations in the manner in which this experience is apprehended and expressed. One of the major tasks of theology is to assess the extent to which the apparently conflicting claims of the world's religions can be understood as due to different culturally conditioned responses to the same reality, or the extent to which their seemingly canceling character indicates their reference is to fantasy rather than fact.

One of the strongest indicators of the validity of the claim that religion is in touch with reality is provided in the universal character of mystical experience, understood as the experience of unity with the ground of all being. It is to be found as an element in all the world's religions. William James concluded his survey of it by saying:

---

[5] The preceding section is from *One World* by John Polkinghorne © 1986 by John Polkinghorne, revised edition published by Templeton Press, 2007, 32–4.

This overcoming of the usual barriers between the individual and the Absolute is the great mystic achievement. In mystic states we become one with the Absolute and we become aware of our oneness. This is the everlasting and triumphant mystical tradition, hardly altered by differences of clime or creed. In Hinduism, in Neoplatonism, in Sufism, in Christian mysticism, in Whitmanism, we find the same recurring note, so there is about mystical utterances an eternal unanimity which ought to make a critic stop and think, and which brings it about that the mystical classics have, as has been said, neither birthday nor native land. Perpetually telling the unity of man with God, their speech antedates language, and they do not grow old.[6,7]

* * *

In science when we discuss physical systems, we are concerned with objects that in some sense we transcend. Pascal said, 'Man is only a reed, the weakest thing in nature; but he is a thinking reed.'[8] That power of thought is what enables us to master and understand the physical world, putting it to the experimental test.

In theology, we are concerned with One who transcends us. There is a mystery in the nature of the Infinite which will never be comprehended by the finite. In theology, there is a tradition called apophatic which acknowledges the essential ineffability of God, the One who is to be met only in clouds and thick darkness. This tradition has been stronger in the contemplative Eastern Orthodox Church than in the rationally confident Latin Church of the West. It is an essential corrective to all theological endeavors, though taken in isolation it would subvert all such endeavors.

God is on the one hand unknowable, but on the other hand, he has acted to make himself known. It is this latter fact that gives theology its mandate. Yet it will always be limited in its power to comprehend. Here is the explanation of the striking contrast in the relative successfulness of science and theology; the one advancing to greater understanding, the other continually wrestling with age-old problems.

The second thing to say is that the inevitable mystery in the nature of God is not a license for irrational assertion about him. Reason has its limitations, but it is not to be trifled with. Paradox may be forced upon

---

[6] W. James, *The Varieties of Religious Experience* (London: Collins, 1977), 404.

[7] The preceding section is from *One World* by John Polkinghorne © 1986 by John Polkinghorne, revised edition published by Templeton Press, 2007, 35–6.

[8] B. Pascal, *Pensées* (New York: Penguin, 1966), 95.

us, our logic and our imaginations may be inadequate to complete the theological task, but we are only to embrace polarities which are required by experience. The insights of faith may not be demonstrable, or even wholly reconcilable, but they are not unmotivated.

The tension between God's eternal nature and his involvement with the world, his timeless knowledge and acts of human choice, is closely allied to an important element in human religious experience. Writing to the Philippians, Paul exhorts them to 'work out your own salvation with fear and trembling; for God is at work in you.'[9] These words reflect a dialectical element in Christian experience: the paradoxical conviction both that we are responsible for what we do and also that it is God who is at work in our lives, molding and transforming them.[10] Before a decision it is the former that dominates our thinking; in retrospect it is the latter that we wish to acknowledge.

This confrontation of responsibility and grace is the human counterpart in our relation to God of the paradoxes we have been considering in his relation to us, his eternal purposes dependent for their fulfillment on the contingent response of men.

Theology and science differ greatly in the nature of the subject of their concern. Yet each is attempting to understand aspects of the way the world is. There are, therefore, important points of kinship between the two disciplines. They are not chalk and cheese, irrational assertion compared with reasonable investigation, as the caricature account would have it. The degree of their relationship is expressed by Carnes when he writes, 'The activities of the theologian are as fallible and his theories as corrigible, as those of any other scientist and any other theories!'[11] He goes on to consider four 'metatheological desiderata;' that is to say, qualities which should characterize the theological enterprise if it is to claim intellectual respectability. They are:[12]

1 *Coherence.* The discourse must hang together. The ultimate achievement of this would be total consistency, but because of the considerations we have been discussing, theology may have to be content to live with some degree of paradox (just as science had to live for a while with the

---

[9] Philippians 2.12–13.

[10] D. M. Baillie believed this to be the fundamental Christian paradox and upon it based his christology: *God Was In Christ* (London: Faber, 1956).

[11] J. R. Carnes, *Axiomatics and Dogmatics* (New York: Oxford University Press, 1982), 68.

[12] Ibid., ch. 5.

unresolved conflict between the wave and particle natures of light until it found the higher rationality of quantum field theory).[13]

2 *Economy*. Theology is not wantonly to multiply entities and explanations. This criterion might be thought to give preference to monotheism over polytheism.

3 *Adequacy*. Theology must be sufficiently rich in concepts to be able to discuss all its matters of concern.

4 *Existential relevance*. There must be an interpretative scheme which links theology with the actual content of religious experience.

Clearly there is a great deal here which is analogous to the demands made of a successful scientific theory.[14]

* * *

If theology is to be found with science in the same spectrum of rational inquiry, it cannot be bracketed off as a 'normative and non-empirical discipline,' as the sociologist Peter Berger describes it.[15] Admittedly theology does not offer predictions open to straightforward empirical testing such as physical science habitually does, at least at its lower levels. Partly that is because theology's understandings are framed in the broad terms of a worldview (so that the sciences most akin to it are those of a historico-observational character, such as cosmology or evolutionary biology, which also does not trade in detailed predictions). Partly it is because God is not to be put to the test (Deuteronomy 6.16). Yet if theology is to maintain cognitive claims, it must be an empirical discipline to the extent that its assertions are related to an understanding of experience. Theological statements unrelated to experience would be arid dogmatism; expressivist utterance not submitted to assessment and interpretation would be mere emotionalism.

As with every other specific form of inquiry, the experience considered by theology will have its own particular character, and discourse about it will need to employ its own particular modes of expression. Observer and object are linked in a mutual relationship. The nature of the object controls what can be known about it and the way in which that knowledge must be expressed. Quite the contrary to what Kant supposed, we do not impose

---

[13] J. C. Polkinghorne, *The Particle Play* (W. H. Freeman, 1979), ch. 5.

[14] The preceding section is from *One World* by John Polkinghorne © 1986 by John Polkinghorne, revised edition published by Templeton Press, 2007, 42–4.

[15] Peter Berger, in R. Gill (ed.), *Theology and Sociology* (New York: Continuum, 1987), 94.

a grid of expectation upon experience as the a priori necessity for its accessibility to us, but our manner of knowing has to be conformed to that with which we have to deal. The quantum world can be investigated and understood, but only on its own counterintuitive and unpicturable terms. A kind of version of the hermeneutic circle is involved. Analysis shows that the uncertainty principle only holds if we commit ourselves wholeheartedly to its consistent application to all phenomena. Otherwise it could be breached, as Einstein tried to do in his long-running battle with Niels Bohr.[16] It would be surprising indeed if the exploration of divine reality were not subject to an analogous need to respect the nature of that encountered.

Theology's necessary conformation to its Object has been expressed by Torrance:

> How God can be known must be determined from first to last by the way in which he actually is known. It is because the nature of what is known, as well as the nature of the knower, determines how it can be known, that only when it actually is known are we in a position to inquire how it can be known.

He goes on to make a comparison with quantum theory and concludes that

> all the way through theological inquiry we must operate with an *open* epistemology in which we allow the way of our knowing to be clarified and modified *pari passu* with advance in deeper and fuller knowledge of the object.[17]

Failure to acknowledge this is the opposite of rationality. God is an altogether different kind of being from any finite existent and he stands in a different relationship to his creation than does any creature contained within it. Basil Mitchell is right to protest that to deny God's existence 'on the sole ground that if he existed he would constitute an exception to the manner in which we normally provide identifying references is to beg the question against the theist by demanding that theism accommodate itself to an essentially atheistic metaphysic.'[18] The One who is the ground of all is bound to be more elusive in his omnipresence that contingent beings of whom one can say 'Lo here' or 'Lo there.'

---

[16] See also N. Bohr, *Atomic Physics and Human Knowledge* (London: Wiley, 1958), 32–66; J. C. Polkinghorne, *The Quantum World* (Longman, 1984), ch. 5.

[17] T. Torrance, *Theological Science* (Oxford: Oxford University Press, 1969), 9–10.

[18] B. Mitchell, *The Justification of Religious Belief* (London: Macmillan, 1973), 19.

In consequence of divine uniqueness, theological language may well be, as Ian Ramsey said, 'object language that exhibits logical peculiarity, logical impropriety'[19] when compared with ordinary usage. It will be strained and stretched as the finite struggles to come to terms with the Infinite. There will be nothing cut and dried about it, for 'theology demands and thrives on a diversity of models – theological discourse must never be uniformly flat. Eccentricity, logical impropriety is its very lifeblood.'[20] Even with such licence for linguistic manoeuvre, the success of theological language will always be strictly limited. 'The religious person must always assert that there can be no formula guaranteed to produce God for inspection,' says Ramsey[21] (any more than a formula would be capable of encapsulating human nature). Theological discourse neither despairs of any utterance (for we are not forced to a total Wittgensteinian silence about that which we do no know) nor does it claim privileged access to otherwise ineffable knowledge. Northrop Frye shrewdly observes that 'it is curious but significant that both "gnostic" and "agnostic" are dirty words in the Christian tradition.'[22]

After taking account of these caveats, we must press on to inquire with Ian Ramsey 'what kind of empirical anchorage have theological words?'[23] His answer was 'a "self awareness" which is more than "body awareness" and not exhausted by spatio-temporal "objects".'[24] For Ramsey, the quintessential experience on which theology is based is a moment of disclosure which reveals a hitherto unseen depth in what is going on. He wove a whole philosophical theology around the discussion of phrases such as 'the penny drops,' 'the ice melts,' indicative of such revelatory experiences.

Of course, the penny might drop in all sorts of ways that would not be reckoned religious, as when a young mathematician first grasps that $\sqrt{2}$ is not rational. Lonergan says that 'Religious conversion is being grasped by ultimate concern. It is other-worldly falling in love.'[25] There is an inescapable element of total personal involvement, so that theological inquiry is not speculative investigation of what might be so but a committed response to what is found to be the case. No doubt Polanyi was right to insist on

---

[19] I. Ramsey, *Religious Language* (London: SCM Press, 1957), 38.

[20] I. Ramsey, *Religion and Science: Conflict and Synthesis* (London: SPCK, 1964), 60.

[21] Ramsey, *Religious Language*, 79.

[22] N. Frye, *The Great Code* (London: Routledge & Kegan Paul, 1982), 67.

[23] Ramsey, *Religious Language*, 14–15.

[24] Ibid.

[25] B. Lonergan, *Method and Theology* (London: Darton, Longman & Todd, 1972), 240.

the intellectual passion involved in doing science, but it pales in comparison with the openness to involvement demanded by the religious quest. Heart and mind must both be engaged for 'Commitment alone without inquiry tends to become fanaticism or narrow dogmatism; reflection alone without commitment tends to become trivial speculations unrelated to real life.'[26, 27]

---

[26] I. G. Barbour, *Myths, Models, and Paradigms* (London: SCM Press, 1974), 136.

[27] The preceding section is from *Reason and Reality* by John Polkinghorne © 1991 by John Polkinghorne, published in the USA by Trinity Press International, 1991, 15–17.

9

Deity

The question of the existence of God is the single most important question we face about the nature of reality. Anthony Kenny says:

> After all, if there is no God, then God is incalculably the greatest single creation of the human imagination. No other creation of the imagination has been so fertile of ideas, so great an inspiration to philosophy, to literature, to painting, sculpture, architecture, and drama; no other creation of the imagination has done so much to stir human beings to deeds of horror and nobility, or set them to lives of austerity or endeavour.[1]

Is that what God is, a gorgeously fertile figment of the imagination, with man the true reality (as Feuerbach believed), or has the thought of him been the inspiration of so much human creativity precisely because he is the creative ground of all that is?

I like the story of a radical English theologian who had been giving a lecture to a group of clergy. At the end, one of them said, 'Professor X, do you believe in God?' He received a carefully nuanced academic reply. 'No, no,' said his interrogator, 'I just want to know if you believe in God.' Professor X then said, 'I believe, indeed I know, that at the heart of reality is One who reigns and loves and forgives.' It was a splendid reply, going with unadorned directness to the heart of what belief in the existence of God is all about. David Pailin says that 'a theistically satisfactory concept of the divine must conceive of the divine as being intrinsically holy, ultimate, personal, and agential.'[2] Does such a concept of God make sense? If so, do we have reason for believing in such a being?

Neither question is easy to answer. God is a different kind of being from any other about which we might speak. He is not part of that metaphysical monism, for his active will is the sustaining ground of that single created reality. Diogenes Allen says, 'God is not the final member of a succession of beings studied in cosmology or in any of the sciences, any

---

[1] A. Kenny, *The Metaphysics of Mind* (Oxford: Oxford University Press, 1989), 121.
[2] D. A. Pailin, *God and the Process of Reality* (Oxford: Oxford University Press, 1989), 24.

more than God was the top storey of Aristotle's hierarchical universe, the unmoved mover.[3] We have to use the language of finite being to try to talk about the infinite – we have no other means at our disposal – but it will always have to be in some stretched, analogical sense. It is less misleading to speak of God as personal rather than impersonal (he is more like a person than a 'force'), but that is not licence for the *naïveté* of the picture of an old-man-in-the-sky.

If the notion of God were incoherent, he could hardly exist. But I am cautious about our powers to assess coherence, particularly when we move into realms of experience far removed from the preserve of everyday thought. In 1900, any competent first-year philosophy undergraduate could have demonstrated the 'incoherence' of anything appearing sometimes like a wave and sometimes like a particle. Yet that is how light was found to behave and, through the universe-assisted logic of that discovery, we have been led to the invention of quantum field theory, which combines wave-like and particle-like behaviour without a taint of paradox.[4] Common sense (and a good deal of philosophy, particularly in the twentieth century, is a kind of painstaking common sense) is not the measure of everything. Thus, though the first, the conceptual, question has logical priority, I attach greater importance to the second, the evidential, question of whether we have reason to believe God actually exists.

These lectures – indeed the whole sequence of Gifford Lectures stretching back to 1888 – are attempts to address that issue. The paradox lies in the fact that He who is most real is also He who is most elusive. Usually we can gain some purchase on the question of the existence of entities by comparing instances of their claimed presence with instances of their acknowledged absence – or, if like gravity, the entity is always present, at least instances of variation in the strength of its effects. But God is always present and he neither waxes nor wanes. The One who is the ground of all that is must be compatible with all that is.

The nearest analogy in the physical world would be a universal medium, such as the nineteenth-century aether or the twentieth-century quantum vacuum. The former faded away when relativity theory left it no work to do; the latter provides a basis for understanding certain pervasive effects (vacuum polarization) and even (if extreme cosmological speculation be true) the origin of this particular universe that we experience (though the

---

[3] D. Allen, *Christian Belief in the Modern World* (Louisville, Ky.: Westminster/John Knox Press, 1989), 74–5.

[4] See J. C. Polkinghorne, *The Particle Play* (W. H. Freeman, 1979), ch. 5.

vacuum itself remains unexplained). The atheist will think God is like the aether. I think the better analogy is with the quantum vacuum. . . . If God is personal, then his presence, though unfailing, will not have the dreary uniformity of the action of a force; but it will manifest itself in ways that are appropriate to the individuality of circumstance. Hence we shall investigate the testimony of revelation, understood, in a way which I hope Lord Gifford would be able to accept, as reference to events or people particularly transparent to the divine presence, rather than understood as some mysteriously endorsed knowledge of what otherwise would be ineffable.

Before we tackle that task, we should not scorn considering philosophy's aid to conceptual clarity. I do not deny its utility. I only protest against any claim to prejudge the issues of experience.

It is a difficult area for an amateur like myself to wander into. The technical discussion often attempts a hard clarity where one might think that more employment of chiaroscuro was called for. It might be useful to recall the remark of Niels Bohr that there are 'two sorts of truth: trivialities, where opposites are obviously absurd, and profound truths, recognized by the fact that the opposite is also a profound truth.'[5] In the latter case, we must strive for coherence by seeking to delimit the range of applicability of the contrasting concepts (as is the case with complementary notions in quantum theory, from which Bohr drew his insight).

It is part of the classical Christian tradition, stemming particularly from St Thomas Aquinas, to lay stress on the simplicity of the divine nature. Naturally this does not mean a facile rational transparency – Pailin rightly says that 'a "simple" theistic faith is a contradiction in terms. God is not that kind of an object'[6] – but an unanalysable unity of being. David Burrell interprets this unity as required, because if there were divine parts they might seem formal causes of the divine nature, which is contradictory in a Being with aseity (being-in-itself).[7]

Anyway, do not Christians, with Jews and Muslims, proclaim their faith in One God? Yet the main import of that proclamation is surely to assert that there is one prevailing will behind the world's existence and so to free us from the ambiguity of a dualism of light and darkness. It is not to make a metaphysical point about the divine nature. An unrelenting emphasis

[5] A. L. Mackay, *The Harvest of the Quiet Eye* (Bristol: Institute of Physics, 1977), 21.

[6] Pailin, *God and the Process of Reality*, 41.

[7] D. Burrell, *Knowing the Unknowable God* (Notre Dame, Ind.: University of Notre Dame Press, 1986), 40.

on divine simplicity easily translates into an image of static undifferentiated perfection, which is inhospitable to the use of personal language about God as Father. How does such an unchanging Unity relate to his changing creation? If we are forbidden to discriminate God's will from God's knowledge, how are we to understand those sinful acts, of whose occurrence he must have perfect knowledge but which are surely contrary to his perfect will?

In our human experience we are aware that the unity of a person is not incompatible with the existence of parts within the psyche, as the insights of modern depth psychology, brilliantly anticipated by the introspective genius of St Augustine, make clear to us. The Christian doctrine of the divine Trinity is exciting, not only as doing justice to the varied economy of our experience of God, but also as the hint of how that property of structured-unity, encountered in ourselves, might extend into the ineffability of divine essence, neither confusing the Persons nor dividing the Substance.

Perichorēsis, the mutual indwelling of the divine Persons which retains unity in diversity, might without impiety be considered under the metaphor of the 'divine bootstrap,' the self-sustaining exchange constituting the aseity of the God whose quintessential nature is love. Something more than the banality of a scientist's metaphor is involved here. Although bootstrap ideas have not proved in the end to be successful in physics, the concept offers a possibility, unenvisaged by classical logicians, of declining to differentiate cause from effect, and so meeting the objection cited above from Burrell. Ward says of Thomist thought that 'its basic error is in supposing that God is logically simple – simple not just in the sense that his being is indivisible, but in the much stronger sense that what is true of any part of God is true of the whole. It is quite coherent, however, to suppose that God, while indivisible, is internally complex.'[8]

I have already spoken several times of God's aseity, that his essence implies his existence, so that his nature is that of a necessarily existing being, needing no explanation in terms of anything else. Concepts of divine necessity can function in a variety of ways. The most straightforward manner is as the ultimate answer to Leibniz's great question, 'Why is there something rather than nothing?' Every chain of explanation has to have a starting-point which is necessary to it, in the sense of its being the unexplained, without the assumption of which it would be impossible to frame an explanation at all.

---

[8] K. Ward, *Rational Theology and the Creativity of God* (Malden, Mass.: Blackwell, 1982), 216.

Intellectually, it is true that nothing comes of nothing. God can play that foundational role for a believer, but for the atheist it would be natural to follow Hume and take the existence of the physical world, with its intrinsic properties, as the ground of explanation. To do so, however, would not be to treat the physical world as necessary in quite the sense in which many theologians want to treat God as necessary. To approximate to the latter, one would have to claim that matter was somehow sufficiently self-explanatory. In fact the physical universe, by its very rational order and fruitfulness, seems to many to point beyond itself, so that there is more intellectual satisfaction in attributing its existence to the will of a self-sufficient agent than in treating it as a fundamental brute fact.

There is a tendency today among some fundamental physicists to believe that there is a unique Theory of Everything (a TOE, as they lightheartedly say) whose discovery is just around the corner and which will then somehow explain why the world is. I very much doubt that ultimate explanation will prove attainable, both because the history of science discourages the expectation of such final achievement (in the past there has always been a surprise waiting around the next corner) and also because those who hope for a TOE have already tacitly decided that it must incorporate quantum theory and gravity, which are empirically necessary but not logically necessary requirements.

Even if I am wrong about finality, a TOE will be more like the precise statement of Leibniz's question, rather than its answer. The most that physical theory could achieve (either in a modern TOE or in the old-style bootstrap programme) is a self-consistency and not a self-sufficiency. We would still want to ask, Why are there things which work in this particular way? In Stephen Hawking's words, 'Even if there is only one possible unified theory, it is just a set of rules and equations. What is it that breathes fire into the equations and makes a universe for them to describe?'[9] In the final paragraph of his book, Hawking seems to me to backtrack and suggest that possession of a TOE might somehow facilitate the discussion of why the universe exists.[10] I think that to suppose that is just a category mistake.

Physics influences metaphysics in various ways, but it is not the same as metaphysics, and Leibniz's question is the ultimate *metaphysical*

---

[9] S. W. Hawking, *A Brief History of Time* (New York: Bantam, 1988), 174.

[10] Ibid., 175. Steven Weinberg, *The First Three Minutes: A Modern View of the Origin of the Universe* (New York: Basic Books, 1993) makes the interesting suggestion that a final theory, though not logically necessary, might be 'logically isolated' so that 'there is no way to modify it by a small amount without the theory leading to logical absurdities' (p. 189).

question. God's will can be its satisfying answer because the multivalued nature of the world's reality, in its order, beauty, ethical imperative and experience of worship, reflects the personal character of a Creator who is rational, joyful, good and holy. The plain assertion of the world's existence leaves unresolved the issue of how our diverse kinds of experience relate to each other. The strategy of materialist atheists is usually to claim that science is all, and that beauty and the rest are merely human constructs arising from the hard-wiring of our brains. I cannot accept so grotesquely impoverished a view of reality. Theism explains much more than a reductionist atheism can ever address. It is worth noting also that, as Talcott Parsons put it, in the history of humanity 'Religion is as much a human universal as language.'[11] Modern Western unbelief has something of the air of a cultural aberration in its rejection of a spiritual dimension to reality.[12]

---

[11] Quoted in J. Hick, *An Interpretation of Religion: Human Responses to the Transcendent* (New York: Macmillan, 1989), 21.

[12] The preceding section is from *The Faith of a Physicist* by John Polkinghorne © 1994 by John Polkinghorne, published in the USA by Princeton University Press, reprinted by permission of Princeton University Press, 52–7.

10

Natural theology

If [theology] is to lay claim again to its medieval title of the Queen of the Sciences, that will not be because it is in a position to prescribe the answers to the questions discussed by other disciplines. Rather, it will be because it must avail itself of their answers in the conduct of its own inquiry, thereby setting them within the most profound context available. Theology's regal status lies in its commitment to seek the deepest possible level of understanding. In the course of that endeavor, it needs to take into account all other forms of knowledge, while in no way attempting to assert hegemony over them. A theological view of the world is a total view of the world. Every form of human understanding must make its contribution to it. The offering of the physical sciences to that end must be made, at least partly, by those who work in them. Theology cannot just be left to the theologians, as is made clear by the recent spectacle of a distinguished theologian writing more than 300 pages on God in creation with only an occasional and cursory reference to scientific insight.[1]

It is as idle to suppose that one can satisfactorily speak about the doctrine of creation without taking into account the actual nature of the world, as it would be to think that the significance of the world could be exhaustively conveyed in the scientific description of its physical processes. There must be a degree of consonance between the assertions of science and theology if the latter are to make sense. Hence there is an urgent need for dialogue between the two disciplines. The arena for their interaction is natural theology.

Natural theology may be defined as the search for the knowledge of God by the exercise of reason and the inspection of the world. There are, of course, those who would deny the possibility of such knowledge. They are by no means all of an atheist or agnostic persuasion. People of religious belief have sometimes been so impressed by the transcendent otherness of God that they have asserted that He is only to be encountered in His

---

[1] J. Moltmann, *God in Creation: An Ecological Doctrine of Creation* (London: SCM Press, 1985).

gracious and specific acts of self-disclosure. He can condescend to us, but we are powerless to reach out to Him. The leading proponent of this point of view in our century has been Karl Barth, who wrote of the God of whom the Christian creeds speak:

> He cannot be known by the powers of human knowledge, but is apprehensible and apprehended solely because of His own freedom, decision and action. What man can know by his own power according to the measure of his natural powers, his understanding, his feeling, will be at most something like a supreme being, an absolute nature, the idea of an utterly free power, of a being towering over everything. This absolute and supreme being, the ultimate and most profound, this 'thing in itself,' has nothing to do with God.[2]

That 'nothing' seems like something of an overstatement. We can acknowledge that natural theology, whose source of insight is by definition limited to the generalities of experience, will not tell us all about God that is humanly accessible. The individual encounter with Him, both our own and that of the spiritual masters preserved in the tradition, will surely be of the highest importance. Yet the world is not just a neutral theater in which these individual revelatory acts take place. Rather, it is itself, if theism is true, the creation of God and so potentially a vehicle also for His self-disclosure. God is to be found in the general as well as in the particular.

Natural theology may only be able to help us to discern 'something like a supreme being, an absolute nature,' and it is certainly powerless by itself to bring us to know the God and Father of our Lord Jesus Christ, but its insights are not for that reason to be despised. There is a great deal more to the structure of matter than chemistry can tell us, with its talk of 92 elements, but it would be foolish to refuse its assistance in an inquiry into what the physical world is made of. Similarly, natural theology can provide valuable help in an inquiry about whether the process of the world is the carrier of significance and the expression of purpose. This role is of special relevance today when so many people find it difficult to see theism as a credible and coherent possibility. Natural theology may be for them a necessary starting point.[3]

* * *

[2] K. Barth, *Dogmatics in Outline* (London: SCM Press, 1949), 23.
[3] The preceding section is from *Science and Creation* by John Polkinghorne © 1988 by John Polkinghorne, revised edition published by Templeton Press, 2006, 7–9.

What does it mean to believe in God today? Different religious communities propose different answers to that fundamental question. I speak from within the Christian tradition, though much of what I say would, I believe, find endorsement from my Jewish and Islamic friends. For me, the fundamental content of belief in God is that there is a Mind and a Purpose behind the history of the universe and that the One whose veiled presence is intimated in this way is worthy of worship and the ground of hope. Here, I sketch some of the considerations that persuade me that this is the case.

The world is not full of items stamped 'made by God' – the Creator is more subtle than that – but there are two locations where general hints of the divine presence might be expected to be seen most clearly. One is the vast cosmos itself, with its 15-billion-year history of evolving development following the big bang. The other is the 'thinking reed' of humanity, so insignificant in physical scale but, as Pascal said, superior to all the stars because it alone knows them and itself. The universe and the means by which that universe has become marvellously self-aware – these are the centres of our enquiry.

Those who work in fundamental physics encounter a world whose large-scale structure (as described by cosmology) and small-scale process (as described by quantum theory) are alike characterised by a wonderful order that is expressible in concise and elegant mathematical terms. The distinguished theoretical physicist Paul Dirac, who was not a conventionally religious man, was once asked what was his fundamental belief. He strode to a blackboard and wrote that the laws of nature should be expressed in beautiful equations. It was a fitting affirmation by one whose fundamental discoveries had all come from his dedicated pursuit of mathematical beauty. This use of abstract mathematics as a technique of physical discovery points to a very deep fact about the nature of the universe that we inhabit, and to the remarkable conformity of our human minds to its patterning. We live in a world whose physical fabric is endowed with transparent rational beauty.

Attempts have been made to explain away this fact. No one would deny, of course, that evolutionary necessity will have moulded our ability for thinking in ways that will ensure its adequacy for understanding the world around us, at least to the extent that is demanded by pressures for survival. Yet our surplus intellectual capacity, enabling us to comprehend the microworld of quarks and gluons and the macroworld of big bang cosmology, is on such a scale that it beggars belief that this is simply a fortunate by-product of the struggle for life. Remember that Sherlock Holmes told a shocked Dr Watson that he didn't care whether the Earth went round the Sun or vice versa, for it had no relevance to the pursuits of his daily life!

Even less plausible, in my view, is the claim sometimes advanced that human beings happen to like mathematical reasoning and so they manipulate their account of physical process into pleasing mathematical shapes.[4] Nature is not so plastic as to be subject to our whim in this way. In 1907, Einstein had what he called 'the happiest thought of my life,' when he recognised the principle of equivalence, which implied that all entities would move in the same way in a gravitational field. This universality of effect meant that gravity could be expressed as a property of space-time itself; physics could be turned into geometry.

Einstein then embarked on a search for a beautiful equation that would determine the relevant geometrical structure. It took him eight years to find it, culminating in the discovery of the theory of general relativity in November 1915. It was a truly beautiful theory but now came the moment of truth. On 18 November, Einstein calculated the prediction made by his theory for the motion of the planet Mercury. He found that it precisely explained a discrepancy in relation to Newton's theory that had baffled astronomers for more than sixty years. Einstein's biographer, Abram Pais, says, 'This discovery was, I believe, by far the strongest emotional experience in Einstein's scientific life, perhaps in all his life. Nature had spoken to him.' Whilst the great man himself said, 'For a few days, I was beside myself with joyous excitement.'[5] It was a great triumph but, if the answer had not come out right, the aesthetic power of the equations of general relativity would have been quite unable in itself to save them from abandonment. It was indeed *nature* that had spoken.

There is no a priori reason why beautiful equations should prove to be the clue to understanding nature; why fundamental physics should be possible; why our minds should have such ready access to the deep structure of the universe. It is a contingent fact that this is true of us and of our world, but it does not seem sufficient simply to regard it as a happy accident. Surely it is a significant insight into the nature of reality. I believe that Dirac and Einstein, in making their great discoveries, were participating in an encounter with the divine. It has become common coinage with contemporary writers about science to invoke, in addressing the general

---

[4] A. Pickering, *Constructing Quarks* (Edinburgh: Edinburgh University Press, 1984), 413.
[5] A. Pais, *Subtle Is the Lord* (Oxford: Oxford University Press, 1982), 253.

public, the idea of a reading of the Mind of God.[6] It is a small, but significant, sign of the human longing for God that apparently this language helps to sell books. There is much more to the Mind of God than physics will ever disclose, but this usage is not misleading, for I believe that the rational beauty of the cosmos indeed reflects the Mind that holds it in being. The 'unreasonable effectiveness of mathematics' in uncovering the structure of the physical world (to use Eugene Wigner's pregnant phrase) is a hint of the presence of the Creator, given to us creatures who are made in the divine image. I do not present this conclusion as a logical demonstration – we are in a realm of metaphysical discourse where such certainty is not available either to believer or to unbeliever – but I do present it as a coherent and intellectually satisfying understanding.

So much for signs of Mind. Where are we to look for signs of Purpose? Before 1859, the answer would have been obvious: in the marvellous adaptation of life to its environment. Charles Darwin, by the publication of *The Origin of Species*, presented us with natural selection as a patient process by which such marvels of 'design' could come about, without the intervening purpose of a Designer being at work to bring them into being. At a stroke, one of the most powerful and seemingly convincing arguments for belief in God had been found to be fatally flawed. Darwin had done what Hume and Kant with their philosophical arguments had failed to achieve, abolishing the time-honoured form of the argument from design by exhibiting an apparently adequate alternative explanation.

Since then, two important developments have taken place. One is the realization in the late 1920s that the universe itself has had a history and that notions of evolving complexity apply not only to life on Earth, but to the whole physical cosmos. The other is the acknowledgement that when we take this cosmic history into our reckoning, evolution by itself is not sufficient to account for the fruitfulness of the world. Let me explain.

A convenient slogan-encapsulation of the idea of evolution is to speak of it as resulting from the interplay of chance and necessity. 'Chance' stands for the particular contingencies of historical happening. This particular cosmic ripple led to the subsequent condensation of this particular group of galaxies; this particular genetic mutation turned the stream of life in this particular direction rather than another. 'Necessity' stands for the lawfully regular environment in which evolution takes place. Without a

---

[6] See, e. g., P. C. W. Davies, *The Mind of God* (Chicago: Simon & Schuster, 1992); S. W. Hawking, *A Brief History of Time* (New York: Bantam, 1988).

law of gravity, galaxies would not condense; without reasonably reliable genetic transmission, species would not be established. What we have come to understand is that if this process is to be fruitful on a cosmic scale, then necessity has to take a very specific, carefully prescribed form. Any old world will not do. Most universes that we can imagine would prove boring and sterile in their development, however long their history were to be subjected to the interplay of chance with their specific form of lawful necessity. It is a particular kind of universe which alone is capable of producing systems of the complexity sufficient to sustain conscious life.

This insight, called the Anthropic Principle, has given rise to much discussion.[7] Is it no more than a simple tautology, saying that this universe which contains ourselves must be compatible with our having appeared within its history? For sure that must be so, but it is surprising – and many of us think significant – that this requirement places so tight a constraint on the physical fabric of our world. Although we know by direct experience this universe alone, there are many other possible worlds that we can visit with our scientific imaginations, and almost all of them, we believe, would be infertile.

John Leslie, who has given a detailed account of the many processes that depend on the precise character of physical law for their ultimately life-generating effect, has also given a careful discussion of what conclusions we might draw from the Anthropic Principle. We are in a realm of discourse where such conclusions depend on the judgement that we have attained a deeper and more comprehensive understanding, rather than that we have deduced a logically unassailable consequence. Leslie believes that it is no more rational to think that no explanation is required of fine anthropic coincidences than it would be to say that my fishing apparatus can accept a fish only exactly 23.2576 inches long and, on casting the rod into the lake, I find that immediately I have a catch, which is simply my good luck – and that's all there is to say about it.[8] The end of the matter for Leslie is this: 'My argument has been that the fine tuning is evidence, genuine evidence, of the following fact: *that God is real, and/or there are many and varied universes.* And it could be tempting to call the fact an observed one. Observed indirectly, but observed none the less.'[9] Either

---

[7] J. D. Barrow and F. J. Tipler, *The Anthropic Cosmological Principle* (Oxford: Oxford University Press, 1986); J. Leslie, *Universes* (London: Routledge, 1989); see also, J. C. Polkinghorne, *Reason and Reality* (SPCK/Trinity Press International, 1991), ch. 6; *Beyond Science* (Cambridge University Press, 1996), ch. 6.

[8] Leslie, *Universes*, 9–13.

[9] Ibid., 198.

there is one world whose fruitful potential is the expression of divine purpose or there are many worlds, one of which just happens to be right for the evolution of life.

Those who wish to avoid any suggestion of a divine purpose manifested in the fruitful fine tuning of physical law will have to opt for the second of Leslie's alternative explanations.[10] There are a variety of ways in which one might conceive of the existence of such a portfolio of different universes, understood as domains in which different laws of nature are operating. The more plausible accounts will seek to make some appeal to scientific knowledge and will not just rely on the ad hoc assumption that there are a lot of separate worlds that just happen to exist.

Many-worlds quantum theory will not do the trick (even if one believed in it, which I do not), for its parallel worlds are simply ones in which quantum events have different specific outcomes and the basic laws of nature are common to them all.[11] Modern ideas about symmetry breaking offer a little more scope. If there is a Grand Unified Theory of the fundamental forces of the universe, then the particular forces that we actually observe, and which are the concern of the Anthropic Principle, will have crystallised out from this highly symmetric ur-state very early in cosmic history, as expansion cooled the world below the relevant transition temperature. The precise details of this symmetry breaking, and the consequent precise force ratios resulting from it, are spontaneously generated through the amplification of tiny random fluctuations. This process need not be literally universal, and the cosmos may be split into vast domains in which different consequences have been realised. The universe observable by us might be a part of one such huge domain, and, of course, in our particular neck of the woods, the force ratios are 'by chance' compatible with our evolution. This account is speculative, but motivated, and I am inclined to consider its possibility as far as it goes. That, however, is not very far. One still needs the right sort of Grand Unified Theory for all this to be feasible, and in that respect our universe is still very special compared to the totality of universes that we can imagine.

Moving up on the scale of bold speculation, one might evoke notions of quantum cosmology which suggest that universes of various kinds are continually appearing as a physical process called inflation blows up

---

[10] A theist could, of course, combine the two options, but personally I find that unappealing.

[11] See, for example, J. C. Polkinghorne, *The Quantum World* (Longman/Princeton University Press, 1984), 67–8; A. Rae, *Quantum Physics: Illusion or Reality?* (Cambridge: Cambridge University Press, 1986), ch. 6.

microworlds, which have bubbled up as quantum fluctuations in some universal substrate.[12] Proponents of this point of view are sometimes moved to describe our anthropic universe as being 'a free lunch.' The phrase itself should trigger a cautious evaluation of the offer being made. The cost of this particular cosmic meal is the provision of quantum mechanics itself (a classical Newtonian world would be a perfectly coherent possibility, but a sterile one), and just the right quantum fields to fluctuate in order to produce first inflation and then all the necessary observed forces of nature. This idea is less well established scientifically than the domain option and, in any case, it does not really remove anthropic particularity, for the basic physical laws still have to take certain specific forms which are the necessary foundation of the proposed quantum cosmology.

Beyond this point, speculation becomes rapidly more rash and more desperate. Maybe, the laws of nature themselves fluctuate, so that a vast portfolio of conceivable, or (to us) inconceivable, worlds rise and fall in the relentless exploration of random possibility – occasional patches of transient and varied order in a sea of seething chaos. We have moved far beyond anything that could be called scientific in this exercise of prodigal conjecture. It is time to consider Leslie's other alternative: that there is a divine purpose behind this fruitful universe, whose 15-billion-year history has turned a ball of energy into the home of saints and scientists, and that this purpose has been at work in just one world of consistent physical law (though maybe with domains of different expressions of that law).

Once again the theistic conclusion is not logically coercive, but it can claim serious consideration as an intellectually satisfying understanding of what would otherwise be unintelligible good fortune. It has certainly struck a number of authors in this way, including some who are innocent of any influence from a conventional religious agenda.[13] Such a reading of the physical world as containing rumours of divine purpose, constitutes a new form of natural theology, to which the insight about intelligibility can also be added. This new natural theology differs from the old-style natural theology of Anselm and Aquinas by refraining from talking about 'proofs' of God's existence and by being content with the more modest role of offering theistic belief as an insightful account of what is going on.

---

[12] The quantum vacuum is an active medium owing to fluctuation effects.

[13] P. C. W. Davies, *God and the New Physics* (London: Dent, 1983); Davies, *Mind of God*; H. Montefiore, *The Probability of God* (London: SCM Press, 1985); J. C. Polkinghorne, *Science and Creation* (SPCK, 1988), chs 1, 2; and n. 4.

It differs from the old-style natural theology of William Paley and others by basing its arguments not upon particular occurrences (the coming-to-be of the eye or of life itself), but on the character of the physical fabric of the world, which is the necessary ground for the possibility of any occurrence (it appeals to cosmic rationality and the anthropic form of the laws of nature).

This shift of focus has two important consequences. The first is that the new-style natural theology in no way seeks to be a rival to scientific explanation but rather it aims to complement that explanation by setting it within a wider and more profound context of understanding. Science rejoices in the rational accessibility of the physical world and uses the laws of nature to explain particular occurrences in cosmic and terrestrial history, but it is unable of itself to offer any reason why these laws take the particular (anthropically fruitful) form that they do, or why we can discover them through mathematical insight. The second consequence of this shift from design through making to design built into the rational potentiality of the universe is that it answers a criticism of the old-style natural theology made so trenchantly by David Hume. He had asserted the unsatisfactoriness of treating God's creative activity as the unseen analogue of visible human craft. The new natural theology is invulnerable to this charge of naive anthropomorphism, for the endowment of matter with anthropic potentiality has human analogy. It is a creative act of a specially divine character.

Physical scientists, conscious of the wonderful order and finely tuned fruitfulness of natural law, have shown significant sympathy with the attitude of the new natural theology. Biological scientists, on the other hand, have been much more reserved. Their attention is focused on the process of the world (particularly, the evolutionary processes of developing terrestrial life) and they pay scant attention to the fundamental physics that underlies that process.[14] They seem to regard it as unproblematic that the chemical raw materials for life are available in our universe. Instead, they look to the variety of life, both in its marvellous fecundity and ingenious strategies for living and also in its wastefulness and suffering, exemplified by the extinction of species and the existence of painful parasitisms. Beneath it all some of them discern no more than the strife of selfish genes struggling for continuing survival. Joy in nature and sorrow

[14] R. Dawkins, *The Blind Watchmaker* (London: Longman, 1986); *River out of Eden* (London: Weidenfeld and Nicolson, 1995).

at its apparent tragedies are alike, to them, vain human musings on the meaningless tale of cosmic history:

> If the universe were just electrons and selfish genes, meaningless tragedies like the crashing of a bus are exactly what we should expect, along with equally meaningless *good* fortune. Such a universe would be neither evil nor good in intention. It would manifest no intentions of any kind. In a universe of blind physical forces and genetic replication, some people are going to get hurt, other people are going to get lucky, and you won't find any rhyme or reason in it, nor any justice.[15]

Whatever this bleak judgement is, it is clearly not a conclusion of science alone. It was not his knowledge of genetics that enabled Richard Dawkins to make this pronouncement. Rather, it represents his metaphysical judgement on the significance of the scientific story which is presented to us. In fact, it is *science* that is 'blind,' for as a self-defining methodological strategy it has closed its eyes to the possibility of discerning evil or good or justice or intention. Those who construct metaphysical theories of wider meanings, or lack of meaning, must take science into account. But there is certainly more than one way in which to do so.[16]

---

[15] Dawkins, *River out of Eden*, 132–3.

[16] The preceding section is from *Belief in God in an Age of Science* by John Polkinghorne © 1998 by Yale University, published in the USA by Yale University Press, reprinted by permission of Yale University Press, 1–12.

11

Creation

In the beginning was the big bang. As the world sprang forth from the fuzzy singularity of its origin, first the spatial order formed, as quantum fluctuations ceased seriously to perturb gravity. Then space boiled, in the rapid expansion of the inflationary era, blowing the universe apart with incredible rapidity in the much less than 10^{-30} seconds that it lasted. The perfect symmetry of the original scheme of things was successively broken as the cooling brought about by expansion crystallized out the forces of nature as we know them today. For a while the universe was a hot soup of quarks and gluons and leptons, but by the time it was one ten-thousandth of a second old, this age of rapid transformations came to a close and the matter of the world took the familiar form of protons and neutrons and electrons. The whole cosmos was still hot enough to be the arena of nuclear reactions, and these continued until just beyond the cosmic age of three minutes. The gross nuclear structure of the universe was then left, as it remains today, at a quarter helium and three-quarters hydrogen. It was far too hot for atoms to form around these nuclei, and this would not occur for another half a million years or so. By then the universe had become cool enough for matter and radiation to separate. The world suddenly became transparent and a universal sea of radiation was left to continue cooling on its own until, 15 billion years later, and by then at a temperature of 3 °K, it would be detected by two radio astronomers working outside Princeton – a lingering echo of those far-off times.

Gravity is the dominant force in the next era of cosmic history. It continued its even-handed battle against the original expansive tendency of the big bang, stopping the universe from becoming too rapidly dilute but failing to bring about an implosive collapse. Although the early universe was almost uniform in its constitution, small fluctuations were present, producing sites at which there was excess matter. The effect of gravity enhanced these irregularities until, in a snowballing effect, the universe after a billion years or so, began to become lumpy and the galaxies and their stars began to form.

Within the stars nuclear reactions started up again, as the contractive force of gravity heated up the stellar cores beyond their ignition temperature. Hydrogen was burned to become helium, and when that fuel was exhausted a delicate chain of nuclear reactions started up, which generated further energy and the heavier elements up to iron. The elemental building blocks of life were beginning to be made. Every atom of carbon in every living being was once inside a star, from whose dead ashes we have all arisen. After a life of 10 billion years or so, stars began to die. Some were so constituted that they did so in the dramatic death-throes of a supernova explosion. Thus the elements they had made were liberated into the wider environment and at the same time the heavier elements beyond iron, inaccessible through the burning of stellar cores, were produced in reactions with the high-energy neutrinos blowing off the outer envelope of the exploding star.

As a second generation of stars and planets condensed, on at least one planet (and perhaps on many) the conditions of chemical composition, temperature and radiation were such that the next new development in cosmic history could take place. A billion years after conditions on Earth became favourable, through biochemical pathways still unknown to us, and utilizing the subtle flexible-stability with which the laws of atomic physics endow the chemistry of carbon, long chain molecules formed with the power of replicating themselves. They rapidly gobbled up the chemical food in the shallow waters of early Earth, and the 3 billion years of the history of life had begun. A genetic code was established, a biochemical alphabet in which the instructions for terrestrial life are universally spelled out. Primitive unicellular entities transformed the atmosphere of Earth from one containing carbon dioxide to one containing oxygen, thereby permitting important developments in metabolism. The process of photosynthesis evolved, *the* method by which the sun's energy is trapped and preserved for the maintenance of all living beings. Eventually, and then with increasing rapidity, life began to complexify through a process which certainly included the sifting of small variations through the environmental pressures of natural selection. Seven hundred million years ago, jellyfish and worms represented the most advanced forms of life. About 350 million years ago, the great step was taken by which some life left the seas and moved on to dry land. Seventy million years ago, the dinosaurs suddenly disappeared, for reasons still a matter of debate, and the little mammals that had been scurrying around at their feet seized their evolutionary chance. Three and a half million years ago, the Australopithecines began to walk erect. Archaic forms of *homo sapiens* appeared a mere 300,000 years ago, and the modern form became established within the last 40,000 years. The universe had become aware of itself.

Such, in outline, is the story that science tells us about the history of the world. There are some speculations (particularly in the very early cosmology) and some ignorances (particularly in relation to the origin of life), but there seems to me to be every reason to take seriously the broad sweep of what we are told. Theological discourse on the doctrine of creation must be consonant with that account.

Of course, the first thing to say about that discourse is that theology is concerned with ontological origin and not with temporal beginning. The idea of creation has no special stake in a datable start to the universe. If Hawking is right, and quantum effects mean that the cosmos as we know it is like a kind of fuzzy spacetime egg, without a singular point at which it all began, that is scientifically very interesting, but theologically insignificant. When he poses the question, 'But if the universe is really completely self-contained, having no boundary, or edge, it would have neither beginning nor end: it would simply be. What place, then, for a creator?'[1] it would be theologically naive to give any answer other than: 'Every place – as the sustainer of the self-contained spacetime egg and as the ordainer of its quantum laws.' God is not a God of the edges, with a vested interest in boundaries. Creation is not something he did 15 billion years ago, but it is something that he is doing now.

An important implication of the Christian doctrine of creation is that it clearly distinguishes the created order from its Creator. Barth says that 'Creation is the freely willed and executed positing of a reality distinct from God.'[2] Burrell says, 'What is at issue here is a clean discrimination of creation from emanation, of intentional activity from necessary bringing forth.'[3] Emanationism pictures the world as arising in a kind of panentheistic way, as the divine being's fruitfulness inevitably spills over into a multiplicity of consequences. In its view, the world is at the hem of deity.

Christian theology, on the contrary, sees the world as the consequence of a free act of divine decision and as separate from deity. The universe's inherent contingency is conventionally and vividly expressed in the idea

[1] S. W. Hawking, *A Brief History of Time* (New York: Bantam, 1988), 141.
[2] C. Green (ed.), *Karl Barth* (London: Collins, 1989), 188.
[3] D. Burrell, *Knowing the Unknowable God* (Notre Dame, Ind.: University of Notre Dame Press, 1986), 15.

of creation *ex nihilo*. Nothing else existed (such as the brute matter and the forms of the classical Greek scheme of things) either to prompt or to constrain the divine creative act. The divine will alone is the source of created being. 'In the doctrine of creation out of nothing, ... Christians replaced the notion of irrational accident or blind chance by the concept of contingence.'[4] God's decision was freely made. This concept can be held to have played an important part in the ideological undergirding of modern science, for it implied both that the world was rational and also that the nature of its rationality depended on the choice of its Creator, so that one must look to see what actual form it had taken.

It is sometimes said that creation *ex nihilo* is just the sort of metaphysical speculation which got grafted on to biblical ideas when Christianity expanded into the late Hellenic world. It is certainly true that it is possible to give a natural exegesis of Genesis 1 which falls short of the explicit articulation of this concept. But I agree with Keith Ward that the doctrine is implicit in the clear claim that all depends upon God's will ('And God said, "Let there be ..."'). 'It is therefore correct to see this doctrine of creation as implicit in the Biblical doctrine that God is the creator of heaven and earth,' says Ward, 'that he can do all things, that nothing is beyond his power.'[5]

The doctrine safeguards the fundamental theological intuition that creation is separate from its Creator, that he has made ontological room for something other than himself. Moltmann says, 'It is only God's withdrawal into himself which gives that *nihil* the space in which God becomes creatively active.'[6] On the other hand, Whitehead rejected the doctrine because he did not want God to play so absolute a role. Whitehead said that God 'is not *before* all creation but *with* all creation.'[7] In their account of process theology, Cobb and Griffin tell us that it 'rejects the notion of *creatio ex nihilo*, if that means creation out of *absolute* nothingness ... Process theology affirms instead a doctrine of creation out of chaos'[8] – which is certainly an exegetically possible view of what is involved in Genesis's reference to that which was 'without form and void' (*tohu wabohu* in Gen. 1.2). But once again I feel that process theology's diminished view of divine power does not allow God to be God.

[4] T. Torrance, *Theological Science* (Oxford: Oxford University Press, 1969), 12.

[5] K. Ward, *Divine Action* (London: Collins, 1990), 6.

[6] J. Moltmann, *The Trinity and the Kingdom of God* (London: SCM Press, 1981), 109.

[7] A. N. Whitehead, *Process and Reality* (New York: Free Press, 1978), 343.

[8] J. B. Cobb and D. R. Griffin, *Process Theology: An Introductory Exposition* (Louisville, Ky.: Westminster Press, 1976), 65.

Needless to say, lighthearted claims that modern physics has provided its own version of creation *ex nihilo* completely miss the point.[9] They are based on speculations about what might have happened in that intrinsically quantum cosmos before the formation of spatial order at the Planck time of 10^{-43} seconds. We need to bear in mind the warning uttered by the great Russian theoretical physicist, Lev Landau, that his cosmologist friends were 'often in error but never in doubt.' All the same, bold speculators are sometimes right, and let us, for the sake of argument, suppose that they are correct in supposing that the universe of our experience has emerged, by one process or another, from a pre-existing quantum vacuum. Only by the greatest abuse of language could such an active and structured medium be called *nihil* (for in quantum theory when there is 'nothing' there, it does not mean that nothing is happening[10]). It is just conceivable that physics may be able to show that given quantum mechanics and a certain gauge field theory of matter, universes will appear; theology is concerned with the Giver of those laws which are the basis of any form of physical reality.

To hold a doctrine of creation *ex nihilo* is to hold that all that is depends, now and always, on the freely exercised will of God. It is certainly not to believe that God started things off by manipulating a curious kind of stuff called 'nothing.' There is no contradiction in holding at the same time a doctrine of *creatio continua*, which affirms a continuing creative interaction of God with the world he holds in being. The two are respectively the transcendent and the immanent poles of divine creativity. Peacocke says, 'The scientific perspective on the world and life as evolving has resuscitated the theme of *creatio continua* and consideration of the interplay of chance and law (necessity) led us to stress the open-ended character of this process of the emergence of new forms.'[11] That would not altogether have surprised St Augustine, who wrote, 'In the first instance, God made everything together without any moments of time intervening, but now He works within the course of time, by which we see the stars move from their rising to their setting.'[12] We do not today take so ready-made a view as Augustine expresses at the start of that passage – though elsewhere he suggested, 'In the beginning were created only germs or causes

[9] See, for example, P. C. W. Davies, *God and the New Physics* (London: Dent, 1983), ch. 16.

[10] See, for example, J. C. Polkinghorne, *The Particle Play* (W. H. Freeman, 1979), 72–5.

[11] A. R. Peacocke, *Creation and the World of Science* (Oxford: Oxford University Press, 1979), 304.

[12] Quoted in E. McMullin (ed.), *Evolution and Creation* (Notre Dame, Ind.: University of Notre Dame Press, 1985), 10.

of the forms of life which afterward developed in gradual course.'[13] Of course, Augustine certainly believed that all was held in being by God's transcendent will: 'the universe will pass away in the twinkling of an eye if God withdraws His ruling hand.'[14,15]

---

[13] Quoted in R. Stannard, *Science and Renewal of Belief* (London: SCM Press, 1982), 11.

[14] Quoted in McMullin, *Evolution and Creation*, 11.

[15] The preceding section is from *The Faith of a Physicist* by John Polkinghorne © 1994 by John Polkinghorne, published in the USA by Princeton University Press, reprinted by permission of Princeton University Press, 71–5.

12

Providence

In the past ten years, there has been a considerable amount of thought and speculation among those concerned with the interface between science and theology, concerning the extent to which it is possible to speak with integrity about the notion of God's acts in the world, whilst at the same time accepting with necessary seriousness what science can say about that world's regular process.[1] Many factors have made this a suitable subject for discussion.

The first is simply that it is a perennial issue on the Christian agenda. The use of personal language about God, however stretched and analogical such language is rightly recognised as being, carries with it the implication of particular divine response to particular creaturely circumstance. God is not like the law of gravity, totally indifferent to context and uniformly unchanging in consequence. The Christian God is not just a deistic upholder of the world. If petitionary prayer, and the insights of a providence at work in human lives and in universal history, are to carry the weight of meaning that they do in Christian tradition and experience, they must not simply be pious ways of speaking about a process from which particular divine activity is in fact absent and in which the divine presence is unexpressed, save for a general letting-be.

Since talk of God is inescapably analogical, talk about God's action has frequently had recourse, in one way or another, to the only form of agency of which we have direct experience, namely our own power to act in the world. I shall make two assumptions about human activity. One is that it is exercised with a certain degree of freedom; that is, our impression of choosing what to do is not an illusion. I am aware, of course, that this

---

[1] I. Barbour, *Religion in an Age of Science* (San Francisco: Harper and Row, 1990), ch. 8; A. R. Peacocke, *Theology for a Scientific Age*, enlarged edn (London: SCM Press, 1993), ch. 9; J. C. Polkinghorne, *Science and Providence* (SPCK, 1989); *Reason and Reality* (SPCK/ Trinity Press International, 1991), ch. 3; *Science and Christian Belief* (SPCK, 1994), published simultaneously as *The Faith of a Physicist* (Princeton University Press, 1994), ch. 4; K. Ward, *Divine Action* (London: Collins, 1990); V. White, *The Fall of a Sparrow* (Exeter: Paternoster Press, 1985).

is a matter of philosophical contention, but I cannot here attempt to enter into that argument. For my present purpose, I shall treat human choice as being an irreducible fact of human experience.

The second assumption I shall make is that we are psychosomatic unities, indivisible animated bodies, and not a dual and separable combination of flesh and spirit. Such a view sits well with our experiences of the inter-dependence of mind and matter (the effect of drugs or brain damage, the execution of willed intentions, our understanding that we have evolved continuously from the original quark soup of the early universe). Needless to say, I cannot solve the problem of how brain and mind relate to each other, but I look for a solution along the lines of a dual-aspect monism, a complementary account of matter in 'information'-bearing-pattern, which I have tentatively and, of course, inadequately discussed on other occasions.[2] Such a stance takes our material constitution seriously. But it does not capitulate to a reductionist materialism, for it asserts with equal vigour the existence of an irreducible mental pole in human nature. Bearing in mind that all conscious knowledge, even of the physical world, is appro-priated mentally, such an even-handed treatment of mind and matter seems absolutely essential if we are to frame a credible account of our experience. That unconscious atoms have combined to give rise to con-scious beings is the most striking example known to us of the hierarchical fruitfulness of our universe, in which there is a nesting and ascending order of being, corresponding to the transitions from physics to biology to psychology to anthropology and sociology.[3]

A further factor of considerable importance is the recognition by twentieth-century science that there are many *intrinsic* unpredictabilities inherent in the process of the physical world. If we define a mechanical system as one whose behaviour is predictable, and so in principle tame and controllable, then our century has seen the death of a merely mechanical universe. Several discoveries have brought this about.

One, of course, is the well-known feature of quantum theory that permits us only to assign probabilities for the observed outcomes of quantum events. Another discovery, relating to effects operating in the macroscopic realm of classical physics and everyday occurrences, is the identification of the

---

[2] J. C. Polkinghorne, *Science and Creation* (SPCK, 1988), ch. 5; *Reason and Reality*, ch. 3; *Science and Christian Belief/Faith of a Physicist*, ch. 1. 'Information' is being used in some highly generalized sense related to dynamic structure, which is beyond my power to specify with precision.

[3] Many writers have commented on this hierarchy. A detailed and itemized discussion is given in Peacocke, *Theology for a Scientific Age*, ch. 12.

widespread sensitivity to minute details of circumstance displayed by those many systems whose behaviour is called 'chaotic.'[4] Since the slightest disturbance totally changes the dynamic behaviour of chaotic systems, caused by the exponential growth of the effects of such perturbations, the theory of chaos describes a realm of intrinsic unpredictability and non-mechanical behaviour.

This latter realisation – that Newtonian physics is not as robust as two and a half centuries of its exploitation had suggested – came as a great surprise. Our minds were unprepared because we had all been bewitched by another great discovery of Sir Isaac: the calculus. This wonderful mathematical method is precisely adapted to the description of continuous and smoothly varying quantities. Its geometrical counterparts are the well-behaved curves we can sketch with our pens upon a sheet of paper. While there are indeed such bland mathematical entities present in the patterns of the world, there are also many entities of a much more jagged character. These are the celebrated fractals, exhibiting roughly the same character on every scale of investigation, saw edges whose teeth are saw-edged, and so on down in an unending proliferation of structure that never settles to a tame unbroken line. Our mathematical imaginations have been greatly enlarged and enriched by this considerably expanded portfolio of possible behaviour. The world is stranger than Newton had enabled us to think.

If a clockwork universe is no longer on the scientific agenda, one must ask what is to take its place? Unpredictability, after all, is an epistemological property, simply telling us that we cannot know in detail the future behaviour of quantum or chaotic systems. Moreover, such behaviour is not totally random. An unstable atom will be able to decay only in certain specific ways and each of these options will have a quantum probability assigned to it, so that in a large collection of atoms of the same kind (a lump of matter), these different future behaviours will occur as calculable fractions of what is happening. A chaotic system is not totally 'chaotic' in the popular sense, corresponding to absolutely random behaviour. Its future options converge to a certain portfolio of possibility called a 'strange attractor' and it is only this limited range of contingencies that will be explored by the system in an apparently haphazard fashion. In consequence, although the detailed future behaviour of a chaotic system is unknowable, there are certain things that can reliably be said about the generic character of what will happen.

---

[4] See J. Gleick, *Chaos* (London: Heinemann, 1988); I. M. Stewart, *Does God Play Dice?* (Oxford: Blackwell, 1989).

There is no logically inevitable way to proceed from epistemology to ontology, from what we can know about entities to what they are actually like. However, unless we believe ourselves to be lost in a Kantian fog – that is, unless we are condemned to groping encounter with phenomena (appearance) and we totally lack any grasp of noumena (reality) – we must suppose there to be some connection between the two. What that connection should be is a central question for philosophy and, perhaps, the central question for the philosophy of science. It can be resolved only by an act of metaphysical decision. Such an act cannot be logically determined a priori, but it can be rationally defended a posteriori, by an appeal to the fruitful success of the strategy adopted.

The decision made by the vast majority of working scientists, consciously or unconsciously, is to opt for critical realism, which one could define as being the attempt to maximize the correlation between epistemological input and ontological belief. In my view, to put the point with extreme brevity, the cumulative success of science provides the necessary support for the pursuit of this strategy.

In the case of the unpredictabilities of quantum theory, this has been the attitude adopted, not only by most physicists but by a great many philosophers as well. Heisenberg's uncertainty principle, which made the epistemological assertion of the simultaneous unknowability of both position and momentum, has been widely interpreted as a principle of indeterminacy, with the ontological implication that quantum entities do not possess at all times definite positions and momenta. The work of David Bohm and his colleagues in framing an alternative quantum ontology, shows clearly enough that this is not a forced move.[5] The extreme popularity of the indeterminacy interpretation has been due, I believe, not just to its chronological priority but also to a certain naturalness about an approach that allows overt epistemology to be the guide of ontological conjecture.

In the case of chaotic systems the same tendency is not apparent. No doubt a historical effect is at work here – after all the subject derived from the study of Newtonian equations, so that a ready-made interpretation was immediately to hand, indeed it is often called 'deterministic chaos.' I shall later argue that what is metaphysical sauce for the quantum goose should be metaphysical sauce for the chaotic gander. At the same time I shall explain how I believe the equations of Newtonian physics should be understood.

---

[5] D. Bohm and B. J. Hiley, *The Undivided Universe* (London: Routledge, 1993); see also J. T. Cushing, *Quantum Mechanics* (Chicago: University of Chicago Press, 1994).

Let us return to the consideration of divine agency. We have seen the theological motivation for speaking of God's action and also something of the character of the physical setting of the world in which such acts would have to take place. It is time to consider what proposals have been under discussion.

A minimalist response is to decline to speak of particular divine actions and to confine theological talk to the single great act of holding the universe in being.[6] Not only is such a timeless deism inadequate to correspond to the religious experiences of prayer and of an intuition of providence, but it is also interesting that it has not commended itself to those scientist-theologians who have written on these matters.[7] They do not suppose that modern science condemns God to so passive a role. Divine upholding of the cosmos whose regular laws are understood as reflections of God's unchanging faithfulness, is part of the story of God's relationship with the unfolding history of creation, but it cannot, and need not, be taken to be the whole of that story.

Much more popular, both as an explicit theory and as a tacit under-standing of what might be involved in providence-talk, has been the idea that God acts only through divine influence on people.[8] It is proposed that it is in the depths of the human psyche, rather than in the process of the external physical world, that divine agency is to be located. God's actions are those of inspiration and encouragement to human persons. A little reflection, however, soon shows that there are grave difficulties with this point of view. First, it implies that God has been an inactive spectator of the universe for most of its history to date, since conscious minds seem not to have been available for interaction with divinity until, at most, the past few million years or so of that 15-billion-year history. Second, and most important, if we take the psychosomatic view of human nature advocated above, God cannot interact with the psyche without also interacting with the physical process of the world, since we are embodied beings. There is no totally separate realm of spiritual encounter, divorced from the physical/mental reality of a dual-aspect monistic world, in which providence can act. God cannot touch our minds without, simultaneously and inextricably, in some way touching our brains as well.

---

[6] G. D. Kaufman, *God the Problem* (Cambridge, Mass.: Harvard University Press, 1972); M. Wiles, *God's Action in the World* (London: SCM Press, 1986).

[7] See the discussion in J. C. Polkinghorne, *Scientists as Theologians* (SPCK, 1996), 31.

[8] E.g. D. Bartholomew, *God of Chance* (London: SCM Press, 1984), 143: divine action 'in the realm of the mind.'

Process theology has sought a way round these difficulties by proposing a view of physical development in which events are the fundamental units, and all events have an experiential aspect that permits divine interaction by way of a 'lure' towards a particular outcome.[9] It would, perhaps, be too crude to characterize this as a panpsychic view of reality, but it certainly seeks to describe an unbroken continuum of processes within which divine interaction with a person or with a proton could both find a place, though obviously at opposite ends of the spectrum. There is then the possibility of providential interaction throughout all cosmic history, with intensification but no qualitative change, at the moment of the arrival of conscious minds on the terrestrial scene.

I have two difficulties with this account of God's activity. One is physical-philosophical: I do not see that the physical world, as disclosed to scientific exploration, can be held to correspond to a concatenation of events in the manner suggested. Quantum physics involves both continuous development (the Schrödinger equation) and occasional sharp discontinuities (measurements) but it does not, to my mind, suggest the discrete 'graini-ness' that process thinking seems to suppose. The second difficulty is theological. The God of process theology works solely through 'persuasion.' There is a divine participation in each event but, in the end, the event itself leads to its own completion. (It is difficult to write about process ideas without a continual lapse into panpsychic-like language.) I think this places God too much at the margins of the world, with a diminished role inadequate to the One who is believed to care providentially for creation and to be its ultimate hope of fulfillment.

An alternative strategy is to exploit rather directly an analogy between God and creation on the one hand and human beings and their bodies on the other.[10] God is then supposed to be embodied in the universe as we are embodied in parts of it, and to act on the whole as we do on the matter that makes us up (in whatever fashion that might be). It seems that many difficulties beset this proposal. First, the universe, though it certainly does not look like a machine, does not look like an organism either. It lacks the degree of coherence and interdependence that charac-terizes the unity of our bodies. To put the matter bluntly, if the world is God's body, where is God's nervous system within it? Second, in our

---

[9] See J. B. Cobb and D. R. Griffin, *Process Theology* (Louisville, Ky.: Westminster Press, 1976). The great exponent of process thought in relation to science and theology has been Ian Barbour.

[10] G. Jantzen, *God's World, God's Body* (London: Darton, Longman and Todd, 1984); for a critique, see Polkinghorne, *Science and Providence*, 18–22.

psychosomatic nature we are constituted by our bodies, and in consequence we are in thrall to them as they change, eventually dying with their decay.[11] The God of Christian theology cannot be similarly in thrall to the radical changes that have taken place within cosmic history and which will continue to happen in the universe's future. Whatever suggestiveness the idea of God's embodiment in the universe might appear to have as a metaphor, it seems that it cannot successfully function as a putative account of divine action.

It is possible, however, to seek to employ the analogical possibilities of relating divine agency to human agency in a more subtle and nuanced manner. When we act, we seem to do so as total beings. It is the 'whole me' that wills the localized action of raising my arm. I am not inclined to think that this is some sort of psychological delusion produced simply by adding together discrete neuron firings in the brain and particular currents in those nerves that cause muscular contractions. On the contrary, it seems plausible that there is a genuine holistic content to human agency. That would imply that there is a top-down action of the whole on the parts, as well as the familiar bottom-up interaction of the parts making up the whole.

The notion of such top-down causality seems to offer an attractive possible analogy to the way in which God could interact with creation.[12] However, it is also important to recognize that, though the notion of top-down causality is motivated by our human experience of agency, it is not by itself an unproblematic or self-explanatory concept. One has to ask the question of how it may be supposed that there is room for the operation of this additional holistic causal principle within the network of physical causality established by the interactions of the bits and pieces making up the whole. In other words, to use a phrase originating with Austin Farrer, we must consider what might be the 'causal joint' connecting the whole to its parts, the human self to its body, God to creation.

Farrer's own answer would be that, at least in relation to providential agency, this is a question we should decline to address, because it is beyond our human power to penetrate the mystery of divine action.[13] He writes in the tradition that speaks of God's primary agency as being at work in and through the secondary agencies of creaturely causality, in an ineffable

---

[11] Christian hope of a destiny beyond death is expressed in terms of God's resurrection act of reconstituting us in our bodily identity in the environment of the new creation.

[12] Peacocke, *Theology for a Scientific Age*, 53–5, 157–60; Polkinghorne, *Science and Christian Belief/Faith of a Physicist*, 77–9.

[13] A. M. Farrer, *Faith and Speculation* (London: A. & C. Black, 1967).

manner which can be affirmed by faith but which is veiled from the pry-
ing eyes of human reason. Despite the venerability of this way of thinking,
sanctioned by St Thomas Aquinas and developed by many subsequent
Christian thinkers, it seems to me to be a fideistic evasion of the problem.
I cannot give up the search for a causal joint, though I certainly acknowledge that our actual attainments in that quest will necessarily be tentative
and provisional. With the nature of human agency still mysterious, we
can hardly dare to aspire to more than hopeful speculation when it comes
to talk of divine agency. Yet the demand for an integrated account of both
theological and scientific insight impels us to the task.

I have said that I do not expect top-down agency to be just a conglomer-
ative effect of a lot of little bits of bottom-up interactions (in the way that
the temperature of a gas is the average of the individual kinetic energies
of its molecules). If holistic causality is present it must be there as a genuine
novelty, and the structure of the relationships between the bits and pieces
must be intrinsic and ontological in character and not just contingent
ignorances of the details of bottom-up process. They must be 'really there'
if they are to provide the causal joint for which we are looking.

Immediately there comes again to mind those widespread unpredictabilities that twentieth-century science has identified as being present in
physical process. If they are to be of significance in relation to holistic
causality, then they must be interpreted, along the lines already discussed,
in a realist way, as being signs of actual ontological openness.

A popular site for such an exploration has been the uncertainties
of quantum events.[14] Because of the almost universal (but not logically
necessary) tendency to give these unpredictabilities an ontological interpretation, it seems as if there is here room for manoeuvre, space for the
operation of a causal joint. The proposal is not, however, without some
difficulties. Subatomic events scarcely look like promising locations for
holistic causality. After all, one could hardly get more 'bits and pieces' than
elementary particles. It is not clear the extent to which the non-locality
of quantum processes modifies that conclusion.[15] Moreover, the 'gaps' of
quantum uncertainty operate only in particular circumstances, namely in

---

[14] W. G. Pollard, *Chance and Providence* (London: Faber & Faber, 1958); articles by N. Murphy
and T. Tracey in R. J. Russell, N. Murphy and A. R. Peacocke (eds), *Chaos and Complexity*
(Vatican: Vatican Observatory, 1995), 289–358. Nancey Murphy is critical of my use of chaos
theory. She demonstrates that epistemology does not entail ontology (no one ever supposed
it did) but she takes unquestioned the indeterministic interpretation of quantum theory,
which depends upon a similar conjecture.

[15] None of the authors cited in the previous note discuss this.

those intermittent events corresponding to acts of measurement. By measurement, we do not mean just observation by a person, but any record of a state of quantum process in the microworld that is obtained by an irreversible registration in the macroworld of everyday occurrence. Acts of agents are located in that same macroworld. In other words, if quantum theory does have a role to play in solving the problem of agency, it will only be because its effects are amplified in some way to produce an openness at the level of classical physics. The continuing perplexities about the quantum measurement problem remind us that we do not fully understand how the levels of the microworld and the macroworld interlock with each other. It does not seem that the proponents of divine action through quantum events have been able to articulate a clear account of how this could actually be conceived as the effective locus of providential interaction.

In these circumstances it seems worthwhile to explore whether there might not also be macroscopic phenomena that would lend themselves to interpretation as possible causal joints. Arthur Peacocke and I have both considered this possibility.

Peacocke's examples have been chosen from cases of dissipative systems far from equilibrium, where small triggers generate large-scale patterns of an impressive kind. Such order out of chaos provides illuminating illustrations of how structures can be formed and maintained when energy is fed into open systems, thus allowing them to swim against the tide of increasing entropy.[16] This is undoubtedly the way in which living systems are able to sustain their form in a world of change and decay. It is not clear, however, that these systems really model top-down agency. First, the character of their order is long-range pattern generated by chains of local correlations and the confining boundary conditions. That seems more sideways than top-down. Second, what is involved by way of consequence is the generation and preservation of structured pattern, whilst agency seems to require a much more open and dynamical exploration of future possibility.

The way a chaotic system traverses its strange attractor seems a more promising model for such open developments, and this has been the basis for my own suggestions.[17] We can consider the many different trajectories

---

[16] Peacocke, *Theology for a Scientific Age*, 53–5; see I. Prigogine and I. Stengers, *Order out of Chaos* (London: Heinemann, 1984).

[17] Polkinghorne, *Science and Providence*, 26–35; *Reason and Reality*, ch. 3; *Science and Christian Belief/Faith of a Physicist*, ch. 1.

through the attractor's phase space (that is, the range of its future possible states) which all correspond to the same total energy. Their different forms are understood as arising from the effects of vanishingly small disturbances that nudge the system along one path or another, the diverging characters of these different paths corresponding to the chaotic system's extreme sensitivity to perturbations.

It is this sensitivity that produces the intrinsic unpredictabilities. In a critical realist re-interpretation of what is going on, these epistemological uncertainties become an ontological openness, permitting us to suppose that a new causal principle may play a role in bringing about future developments. The character of this principle would be two-fold. First, since the paths through the strange attractor all correspond to the same energy, we are not concerned with a new kind of energetic causality. The energy content is unaffected whatever happens. What is different for the different paths through phase space is the unfolding pattern of dynamical development that they represent. The discriminating factor is the structure of their future history, which we can understand as corresponding to different inputs of *information* that specify its character (this way, not that way). Second, although the diagnostic indicator of chaotic systems is their sensitivity to small triggers, rather than this implying that we should consider them at the level where these individual small fluctuations occur, it forces on us, in fact, a holistic treatment, since the systems' vulnerability to disturbance means that they can never be isolated from the impact of their total environment.

Thus a realist reinterpretation of the epistemological unpredictabilities of chaotic systems leads to the hypothesis of an ontological openness within which new causal principles may be held to be operating which determine the pattern of future behaviour and which are of an holistic character. Here we see a *glimmer* of how it might be that we execute our willed intentions and how God exercises providential interaction with creation. As embodied beings, humans may be expected to act both energetically and informationally. As pure spirit, God might be expected to act solely through information input. One could summarise the novel aspect of this proposal by saying that it advocates the idea of a top-down causality at work through 'active information.' This is a phrase that Peacocke uses also. I locate the relevant causal joint in chaotic dynamics; he appears to regard God as constituting the 'boundary condition' of the universe.[18]

---

[18] Peacocke, *Theology for a Scientific Age*, 59–61, 161–5, 203–6. See the discussion of J. C. Polkinghorne, *Scientists as Theologians* (SPCK, 1996), ch. 3.

I shall make a series of comments on this proposal, first of a scientific character and then in relation to theology.

The first scientific comment is whether one could not combine the widely acknowledged exquisite sensitivity of chaotic systems together with the widely believed openness of quantum systems, to yield a theory whose openness would result from the vulnerability of the macroscopic system to the indeterminate details of its microscopic quantum constituents. Putting it another way, macroscopic openness could be chaotically amplified quantum openness. In the end, of course, there must be a unified account combining the microscopic and the macroscopic, since there are not two physical worlds but one world encountered at these two different levels. However, the difficulties we have in understanding fully how the two levels relate to each other makes me wary of claiming an immediate synthesis. Not only is there the unsolved measurement problem to which we have already referred but also there is still considerable perplexity about what correspondence can be established between chaotic dynamics and quantum mechanics. Without attempting a detailed technical discussion, I must content myself simply to note that the nature of the compatibility of the two has not been established.[19]

The second scientific comment is that, if the proposal is correct, then at the macroscopic classical level the Newtonian deterministic equations for bits and pieces are only approximately valid as limiting cases of more subtle and flexible laws of nature in which the behaviour of parts is dependent on the setting of the whole in which they participate. This contextualism is the way in which top-down influence is brought to bear. The limit involved in obtaining the Newtonian description is obviously separability, achieved in those situations (which certainly exist but which are a subset of all possible occurrences) in which a part can effectively be isolated from the context of its whole. These are precisely the situations in which most experimental investigation takes place, since the relevant system must be capable of being treated locally and separated from its cosmic context if we are to be able to understand its behaviour. Experimental science is possible precisely because there are these cases that can be treated piecemeal, without a universal knowledge of all that is. There are many examples, however, which show that this is not universally the case

---

[19] See article by J. Ford in P. C. W. Davies (ed.), *The New Physics* (Cambridge: Cambridge University Press, 1989), 348–72. See also, however, the logic-based discussion of how classical determinism may be considered to emerge from quantum mechanics, given in R. Omnes, *The Interpretation of Quantum Mechanics* (Princeton: Princeton University Press, 1994), esp. 227–34. Omnes regards this emergence as problematic for chaotic systems.

for chaotic systems.[20] In our experiments we are only able to investigate thoroughly a part of what is going on.

It is important to understand what is involved in this proposed reinter-pretation of what is often called deterministic chaos.[21] The original theory had a deterministic ontology (expressed by its Newtonian equations) but this resulted in an unpredictable epistemology. Instead of adopting the conventional strategy of saying that this shows that simple determinism underlies even apparently complex random behavior, I prefer the realist strategy of seeking the closest alignment of ontology and epistemology (theory and behaviour) by modifying the theoretical basis along the lines proposed. This strategy then has the additional advantage of accommodat-ing the notion of top-down causality in a natural way.

I do not doubt that reluctance to embrace the notion of flexible and contextual laws of nature stems from the fact that a theory of this kind has not yet been formulated in any detail, whilst the alternative interpret-ation of 'deterministic chaos' (localized inflexibility with mere epistemo-logical ignorance of determining detail) has the time-honoured equations of classical dynamics as its rigorous articulation.

Recently, however, Ilya Prigogine has produced some ideas that seem to be very helpful in indicating the form that a more holistic and open dynamical theory might take.[22] He studies certain equations, such as the Liouville equation of statistical mechanics, which describe the development in time of dynamical systems. One can first look for integrable solutions of these equations, that is to say solutions which have a smooth, well-behaved character such as we considered earlier when discussing the calculus. These solutions turn out to have the property that they can always be decomposed into sums of contributions from definite trajectories corresponding to specific picturable behaviours of parts of the system being investigated. In other words, smooth mathematical behaviour yields a localized, bits and pieces, physics account of what is happening. It is, however, mathematically possible to enlarge the class of solutions that will be admitted, in order to include what are called non-integrable solu-tions. These are not so mathematically 'nice' and well-behaved – their introduction corresponds to something like a transition from smooth curves to jagged fractals. It turns out that this enlargement of the range

---

[20] See Polkinghorne, *Science and Providence*, 28–9.

[21] I am grateful to Professor R. J. Russell for a helpful conversation on the issues involved.

[22] I. Prigogine, 'Time, Chaos and the Laws of Physics,' a lecture given in London, May 1995. I am grateful to Professor Prigogine for making the text available.

of mathematical imagination produces possible behaviours that cannot be reduced to a sum of localized specific trajectories. A holistic account is then necessary and at the same time a rigid determinism is no longer present. Prigogine says of these additional solutions that 'instead of expressing certitudes, they are associated to "possibilities."'

Here we are presented with a model of how it can be that Newtonian ideas, which work so well for isolable systems, are not the whole story of what is going on. The new wine of chaos theory bursts the mathematical wine skins of continuous function theory. The world is indeed stranger and more exciting than Newton imagined, even at the level of his own splendid achievements.

A final scientific comment relates to the character of causality through active information. The word 'information' is being used in this slogan phrase to represent the influence that brings about the formation of a structured pattern of future dynamical behaviour. This is not the same as the registration or transmission of bits of information in the sense used by telephone engineers or, more formally, by the mathematical theory of communication. A much closer analogue is provided by the 'guiding wave' of Bohm's version of quantum theory. The latter encodes information about the whole environment (it is holistic), and it influences the motion of a quantum entity by directional preferences but not by the transfer of energy (it is active in a non-energetic way). For information in the sense of the telephone engineer, there is a necessary cost in energy input, since the signal has to rise above the level of the noise of the background. For the Bohmian guiding wave there is no such energy tariff; the wave remains effective however greatly it is attenuated. I believe, therefore, that it is possible to maintain a clear distinction between energetic causality and 'informational' causality, in the sense of the model under discussion.[23,24]

* * *

I propose that human beings act in the world through a combination of energetic physical causality and active information, and that God's providential interaction with creation is purely through the top-down input

---

[23] Cf. the discussion of Bohm and Hiley, *Undivided Universe*, 35–8.

[24] The preceding section is from *Belief in God in an Age of Science* by John Polkinghorne © 1998 by Yale University, published in the USA by Yale University Press, reprinted by permission of Yale University Press, 48–67.

of information. Many theological consequences flow from adopting this point of view.

(1) One of the dilemmas of talk about divine agency has been to find a path between the ineffable mystery of the claim presented by the idea of primary causality and the unacceptable reduction of the Creator to an invisible cause among competing creaturely causes (making God just a physical interventionist poking an occasional divine finger into the processes of the universe). The continuous input of active information appears to offer the opportunity of such a tertium quid.[25] It is the translation into the mundane language of conjecture about causal joints, of a long tradition of Christian thinking that refers to the hidden work of the spirit, guiding and enticing the unfolding of continuous creation.

(2) If it is the unpredictabilities of physical process that indicate the regions where forms of holistic causality can be operating, then all such agency, including divine providence, will be hidden within these cloudy domains. There will be an inextricable entanglement – it will not be possible to itemize occurrences, saying that God did this and nature did that. Faith may discern the divine hand at work but it will not be possible to isolate and demonstrate that this is so. In this sense, the causal joint is implicit rather than explicit. The veiled presence of God, discreetly hidden from contact with finite human being, may be held to require divine actions to be thus cloaked from view. The theological assessment of the balance between what God does and what creatures do, is the old problem of the balance between grace and freewill, now being considered on a cosmic scale.

(3) There are, of course, predictable aspects of natural process that the divine consistency can be expected to maintain undisturbed as signs of God's faithfulness. The succession of the seasons and the alterations of day and night will not be set aside.

(4) Considerations of divine consistency lead us to expect that in comparable circumstances God will act in comparable ways, though the infinite variety of the human condition means that no simple lessons can be drawn from this about individual human destinies. In unprecedented circumstances, it is entirely conceivable that God will act in totally novel and unexpected ways. That is how I try to understand claims about divine miracles, a subject which lies outside the humdrum

---

[25] I am grateful to Professor R. J. Russell for a helpful conversation on this point.

limits of the present discourse,[26] but one which is of central importance to a Christian thinker because of the pivotal role played by Christ's resurrection.

(5) If the physical universe is one of true becoming, with the future not yet formed and existing, and if God knows that world in its temporality, then that seems to me to imply that God cannot yet know the future. This is no imperfection in the divine nature, for the future is not yet there to be known. Involved in the act of creation, in the letting-be of the truly other, is not only a kenosis of divine power but also a kenosis of divine knowledge. Omniscience is self-limited by God in the creation of an open world of becoming.[27]

* * *

If this picture of divine agency is right, a number of consequences flow from it. First, divine action will always be hidden, for it will be contained within the cloudiness of unpredictable processes. The sensitivity of these processes implies that the different forms of causality present can never be separately identified and disentangled from one another. One cannot say, 'This event was due to nature,' and 'That event was due to divine providence.' This seems to me to be the appropriate reflection in the physical world of that theological necessity we discussed earlier, that God neither does everything nor does nothing, but God interacts, patiently and lovingly, with the process of creation, to which the Creator had given its own due measure of independence.

This intermingling of providential grace with the freedom of nature means that divine action will not be demonstrable by experiment, though it may be discernible by the intuition of faith. A bystander on the bank of the Red Sea many years ago might have seen what appeared to be no more than a fortunate coincidence, in that a wind arose to drive back the waters and allow a bunch of fleeing slaves across, but one of those slaves might legitimately have seen in that event God's great act of the deliverance of Israel from oppression in Egypt.

Second, though there are many clouds in the world, there are also some clocks. The regularity of the mechanical aspects of nature are to be

---

[26] Cf. Polkinghorne, *Science and Providence*, ch. 4.

[27] The preceding section is from *Belief in God in an Age of Science* by John Polkinghorne © 1998 by Yale University, published in the USA by Yale University Press, reprinted by permission of Yale University Press, 71–3.

understood theologically as signs of the faithfulness of the Creator. God will not overrule them. Long ago, in hot Alexandria, the great Christian thinker Origen acknowledged that it did not make sense to pray for the cool of spring while enduring the heat of summer. The regular succession of the seasons is mechanical, and it will not be set aside.[28]

---

[28] The preceding section is from *Quarks, Chaos, and Christianity* by John Polkinghorne © 1994 by John Polkinghorne, published in the USA by The Crossroad Publishing Company, 2005, 72.

13

Prayer and miracle

The practice of prayer is central to religion. It is no accident that the great spiritual autobiography of Augustine's *Confessions* is cast in the form of an extended prayer. Peter Baelz is right to say, 'Prayer is a touchstone of a man's religious beliefs.'[1] Of course, prayer is a complex activity, with many aspects to it. It includes worship, the acknowledgment of the greatness of God. It includes a meditative waiting upon him in stillness and silence. For those who are far advanced in its practice, it will include the contemplative experience of unity with the divine. But for all, it will also include petition, the asking of something from God, for ourselves or for others. Jesus encourages this in the Gospels with an embarrassing directness. 'Ask, and it will be given you; seek, and you will find; knock, and it will be opened to you.'[2] Petition is the form of prayer that relates directly to the issue we are considering in this essay.

One could hardly imagine oneself asking the God of deism for anything. One might well adore him for his mighty act of creation, but one could not expect him to do anything about individual happenings within its process. The best one could hope for would be that he had so cleverly constituted his timeless action that things would work out reasonably well. Petitionary prayer implies belief in a God who acts in the particular as well as in the general. We have given reasons why, with appropriate safeguards for creaturely freedom, belief in such a God is a coherent possibility.

We move closer to the mark with Augustine's comment that 'God does not need to have our will made known to him – he cannot but know it – but he wishes our desire to be exercised in prayer that we may be able to receive what he is preparing to give.' In other words, prayer is neither the manipulation of God nor just the illumination of our perception, but it is the alignment of our wills with his, the correlation of human desire and divine purpose. That alignment is not just a passive acceptance of

---

[1] P. Baelz, *Prayer and Providence* (London: SCM Press, 1968), 10.
[2] Matthew 7.7.

God's will by human resignation (though 'if it be thy will' is an essential part of any prayer, since God is the necessary partner in it), but it is also a resolute determination to share in the accomplishment of that will (so that prayer is never divorced from action, nor a substitute for it). Prayer is a collaborative personal encounter between man and God, to which both contribute.[3]

* * *

The cooperation with God involved in prayer is not limited to making available our capacity for action. John Lucas makes an extremely important point when he says:

> We are not only, though within limits, the originators of actions, but also, though within limits, the origin of values . . . The mere fact that we want something is a reason, though not a conclusive reason, for God giving it to us . . . By creating us and the world he has abdicated not merely absolute sway over the course of events but also absolute sway over the scale of values.[4]

Here is another reason why we have to ask, to commit ourselves to what it is that we desire. The blind man who comes to Jesus has to declare what it is he wants done for him.[5] The encounter of prayer is genuinely two-way; we are not faced by God with an illusion of choice. He is not a celestial Henry Ford, offering us a car of any color provided it is black.

It is an astonishing thought that our preferences should play a part in determining what is to be achieved through creation, but that is part of the loving respect of a Father for his children. Loving respect is due also from children to their Father. One of the reasons why we must seek the coming of God's Kingdom through our prayer is that thereby, says Vincent Brümmer, 'we acknowledge that his perfect goodness (on which we can count) does not exclude his being a person (upon whose free decision we may not presume).'[6] The necessity for prayer is well summarized by H. D. McDonald when he writes:

> It may indeed be that God does give His best possible to every man without prayer, for He makes his sun to rise on the evil and the good. But the best

---

[3] The preceding section is from *Science and Providence* by John Polkinghorne © 1989 by John Polkinghorne, revised edition published by Templeton Press, 2005, 80.

[4] J. R. Lucas, *Freedom and Grace* (London: SPCK, 1976), 40.

[5] Mark 10.51.

[6] V. Brümmer, *What Are We Doing When We Pray?* (London: SCM Press, 1984), 54.

possible that God, as faithful Creator, assures without prayer to every man may not be the best possible which could come to any man if he really prayed.[7,8]

* * *

Much the bluntest claim that God acts in the world is made by those who assert that they believe in miracle. C. S. Lewis gives a 'crude and popular' definition of miracle as 'an interference with Nature by supernatural power.'[9] He goes on to acknowledge that this is not a theologian's definition. Let us also consider the definition offered by the philosopher of religion, Richard Swinburne, when he says that a miracle is 'an event of an extraordinary kind, brought about by a god, and of religious significance.'[10] An important qualification added by Swinburne is that of significance. A miracle is not just an astonishingly odd event, such as would be the sudden materialization in Trafalgar Square of a twelve-foot-high statue of Nelson made of chocolate. It has also to be the carrier of meaning. In the Johannine language of the New Testament, a miracle must be a 'sign.' The reason is clear. The only miracles that seriously could be said to be on the agenda are not just acts of a 'supernatural power' or 'a god.' They are the acts of God himself. He is no celestial conjurer, doing an occasional turn, but his actions must always be characterized by the deepest possible consistency and rationality. Therefore they must be endowed with meaning and be free from caprice.[11]

* * *

It is with that word 'interference' that the troubles begin. We can imagine an agent of limited ability interfering with the work of another such agent. You construct a clock. I decide to modify its mechanism so that it no longer keeps me awake by striking the quarters at night. But if I am a perfectly skillful clockmaker I shall surely make for myself the perfect clock at my first attempt. God is not a demiurge, struggling to make the best of

---

[7] H. D. McDonald, *The God Who Responds* (Cambridge: James Clarke, 1986), 115–16.

[8] The preceding section is from *Science and Providence* by John Polkinghorne © 1989 by John Polkinghorne, revised edition published by Templeton Press, 2005, 83–4.

[9] C. S. Lewis, *Miracles* (London: Geoffrey Bles, 1947), 15.

[10] R. Swinburne, *The Concept of Miracle* (Basingstoke: Macmillan, 1970), 1.

[11] The preceding section is from *Science and Providence* by John Polkinghorne © 1989 by John Polkinghorne, revised edition published by Templeton Press, 2005, 53.

recalcitrant brute matter. He is the Creator and Sustainer of the whole physical world. Those very laws of nature, said to be violated by a miracle, are themselves the expression of his Creatorly will. One does not doubt, in one sense, his capacity to countermand them. Such action of itself cannot be beyond the power of an omnipotent God.

Sir George Stokes robustly made the point in his Gifford Lectures of 1891, when he said, 'Admit the existence of a God, of a personal God, and the possibility of miracle follows at once. If the laws of nature are carried on in accordance with his will, he who willed them may will their suspension.'[12] Undoubtedly – but will the rationally coherent God actually change his mind? Will he really work against the grain of the natural law that he himself has ordained? And if that is what he does, why does he not do it more often? There seems to be plenty of scope for extra miracles to alleviate the sufferings of mankind. A theologically acceptable account of miracles will have to incorporate them within a total, and totally consistent, understanding of God's activity, and not see them as singular exceptions.

Thus I do not believe that interference is a fitting word to use about God's relation to his creation. The problem of miracle is twofold. One question is the nature of the evidence which might lead us to suppose that any particular event claimed as a miracle had actually happened. Another question is whether extraordinary events of the kind called miraculous can be any part of the faithful action of God. Is he not the God of reliable process and not of magic? Clearly the second question is prior to the first, since if miracle is an absurdity it is certainly not an act which God has actually performed.[13]

* * *

We are familiar in many branches of knowledge with the utility of dividing up what we know at root to be a fundamental unity. Levels of behavior which are always present may be visible only in particular regimes. The laws of nuclear force act all the time and are indispensable in maintaining the stability of matter, yet we are only aware of their operation when we enter a regime of sufficiently high energy where, for instance, nuclear transmutations become possible which are not observable in ordinary

[12] Quoted in E. L. Mascall, *Christian Theology and Natural Science* (London: Longman, 1956), 180.

[13] The preceding section is from *Science and Providence* by John Polkinghorne © 1989 by John Polkinghorne, revised edition published by Templeton Press, 2005, 54–5.

circumstances. Nowhere in the world was there a nucleus with atomic number greater than 92 until the specially contrived circumstances brought about at the Radiation Laboratory at Berkeley permitted the formation of a series of transuranic elements. Sometimes such changes of circumstance can produce radically different modes of behavior. One example, too familiar to surprise us but remarkable nevertheless, is the way in which the slow increase of temperature suddenly produces a discontinuous change from liquid to gas at boiling point. The detailed physics of such phase changes (as they are called) are notoriously difficult to figure out, but certainly the underlying laws of nature do not change at 100 °C.

That example of the discontinuous change of behavior with changing physical regime, coupled with the unbroken regularity of physical law, may be of some small analogical help in thinking how God might be capable of acting in miraculous, radically unexpected, ways, while remaining the Christian God of steadfast faithfulness. That is the fundamental theological problem of miracle: how these strange events can be set within a consistent overall pattern of God's reliable activity; how we can accept them without subscribing to a capricious interventionist God, who is a concept of paganism rather than of Christianity. Miracles must be perceptions of a deeper rationality than that which we encounter in the every day, occasions which make visible a more profound level of divine activity. They are transparent moments in which the Kingdom is found to be manifestly present.[14]

For all its stark contradiction of normal expectation, the resurrection is readily accommodated in Christian theology within such a consistent account of God's action in Christ. It was fitting that he whom uniquely 'God made . . . both Lord and Christ' should be raised up because 'it was not possible for him to be held by [the pangs of death].'[15] Much more difficult is a claimed occurrence like the turning of water into wine at Cana in Galilee. At one level, it seems an over-reaction to a mild social problem arising from inadequate prior provision. At a somewhat deeper level, it is an acted parable of the transforming power of Christ, but performed in a self-conscious way which does not square easily with the hidden and unforced nature of Jesus' ministry. Christians will take different views on this particular question, but it is clear where the debate lies. Mere wonderworking, without an underlying consistency of action and intent, would never be a credible Christian miracle.

---

[14] Matthew 11.2–6.
[15] Acts 2.24, 36.

The concept of regime, of the sensitive relationship of possibility to circumstances, can also help us to understand something of why miracles occur so sparsely and with a seeming fitfulness. If God is consistent he must act in the same way in the same circumstances, but personal matters are so infinitely graded in their characters that what may seem closely similar occasions can in fact be quite different from each other. In one place, Swinburne defines a miracle as 'a non-repeatable exception to the operation of nature's laws, brought about by God.'[16] Clearly the discontinuous language of 'exception' is exactly what we are trying to avoid, and the word 'unrepeatable' has about it that air of arbitrariness which we are at pains to reject. It can be saved from that if we interpret it as referring to that subtle complexity of human circumstance which implies that personal events are never repetitions of their predecessors. Every human experience is unique. Presumably Farmer had something like this in mind when he wrote: 'It is part of the essential personal quality of the awareness of miracle that it should be in any one experience comparatively rare.'[17] Seldom will the circumstances be just right for the emergence of the unexpected. (That remark is saved from mere tautology by its pointing to the ground that permits miracle to happen.) There remains, of course, the very difficult question of why miracle should be so *exceedingly* rare, when we consider the multitude of agonizing occasions which might be thought to call for its assistance. People say that they cannot at all believe in a God who acts if he did not do so to stop the Holocaust. If God were a God who simply interferes at will with his creation, the charge against him would be unanswerable. But if his action is self-limited by a consistent respect for the freedom of his creation (so that he works only within the actual openness of its process) and also by his own utter reliability (so that he excludes the shortcuts of magic) it is not clear that he is to be blamed for not overruling the wickedness of humankind.

Some further answer might lie in the very specific qualities required of a regime if it is to be able to exhibit what we call the miraculous. The Gospels portray one aspect of this when they record that at Nazareth Jesus 'could do no mighty work there ... and he marveled because of their unbelief.'[18] His healings were not just naked acts of power imposed without the collaborative assent of those to be healed. Augustine discussed:

---

[16] R. Swinburne, *Faith and Reason* (Oxford: Oxford University Press, 1981), 186.

[17] H. H. Farmer, *The World and God: The Study of Prayer, Providence, and the Miracle of Christian Experience* (London: Nisbet, 1935), 125.

[18] Mark 6.5–6 (WEB).

'Why, it is asked, do miracles never occur nowadays, such as occurred (you mention) in former times?'[19] He thought, in fact, that some had occurred in his own time (he gives examples), but that they were more frequent in apostolic times, because they were then necessary to launch the Christian gospel, which subsequently could propagate without such aid. C. S. Lewis makes a similar point about the necessary aptness of historical circumstance when he writes chillingly that:

> God does not shake miracles at Nature at random as if from a pepper-castor. They come on great occasions: they are found at the great ganglia of history – not of political or social history, but of that spiritual history which cannot be fully known by men . . . Miracles and martyrdoms tend to bunch together about the same areas of history.[20]

There are those who would interpret this phenomenon in a different way. They would say that miracles 'occurred' at times of particular ignorance and credulity, or occasions when heady excitement suspended sober judgment. Miracles always seem to happen at some other place, and some other time, than here and now. That challenge reminds us of the first of the two general questions we raised earlier. Our discussion so far has sought to show that miracles are neither ruled out by scientific knowledge that the world is a relentlessly inflexible mechanism (it is not) nor by theological knowledge that God is just the deistic upholder of general process (he is more than that). That there may have been miracles is a coherent possibility.[21]

---

[19] Augustine, *The City of God*, trans. H. Bettenson (Harmondsworth: Penguin Books, 1972), XXII, 8.

[20] Lewis, *Miracles*, 201.

[21] The preceding section is from *Science and Providence* by John Polkinghorne © 1989 by John Polkinghorne, revised edition published by Templeton Press, 2005, 59–63.

14

Time

From the era of debate in ancient Greece between the followers of Parmenides and the followers of Heraclitus concerning the contrasting roles of stability and flux in the essential nature of reality, down two and a half millennia to the present day, the true nature of time has been a matter of continuing metaphysical dispute. The modern descendants of Parmenides adopt the stance that is called 'the block universe.' They believe that the true physical reality is the atemporal totality of the spacetime continuum – the whole of history, past, present and future taken together – and that human experience of the flow of time is just a trick of our limited psychological perspective as we trek along those paths through spacetime that the physicists call 'world-lines.' The fact of the matter for these latter-day Parmenidians is that the cosmic narrative is in some sense 'already' written, and human beings are simply laboriously deciphering the text, line by line. The descendants of Heraclitus, on the other hand, believe that we live within the continuously unfolding process of a world of true becoming. The future is not 'up there,' waiting for us to arrive, but we play our part in making it as we participate in the ever-developing history of the universe.

In the arguments between these two parties one encounters, once again, a debate that can be influenced by science, but which can be settled only by philosophical decision.[1] Proponents of the block universe frequently appeal to special relativity in aid of their point of view. That theory is based on assigning a fundamental role to light, understood to have the property of conveying a signal whose velocity is independent of the state of motion of the source emitting it. Of course, this postulate flies in the face of commonsense expectation.[2]

---

[1] C. J. Isham and J. C. Polkinghorne, 'The Debate over the Block Universe' in R. J. Russell, N. Murphy and C. J. Isham (eds), *Quantum Cosmology and the Laws of Nature* (Vatican Observatory, 1993), 135–44; J. C. Polkinghorne, *Faith, Science and Understanding* (SPCK/Yale University Press, 2000), ch. 7.

[2] The preceding section is from *Exploring Reality* by John Polkinghorne © 2005 by Yale University, published in the USA by Yale University Press, reprinted by permission of Yale University Press, 113–14.

The block theorist says that since the same pair of events could be judged either simultaneous or as occurring at different times, depending on who observes the process, it must follow that time differences are not actually significant and so equal reality must be assigned to past, present and future. The temporal theorist disagrees. Any observer's judgement about distant events is always a *retrospective* matter, since there can be no knowledge of such events until a signal is received conveying the information. It is a consequence of relativity theory that when this signal has been received, the event itself is unambiguously in the past (technically, the event then lies within the recipient's past lightcone, and the characterisation of that lightcone is independent of the observer's state of motion). In other words, judgements of simultaneity refer only to how observers organise their descriptions of the unalterable past, and therefore arguments based on such assessments can do nothing to establish the reality of the still-anticipated future.

Another argument sometimes advanced in defence of the block universe is to point to the failure of physics to incorporate into its account of nature any representation of 'now,' the present moment. Since there is no preferred state of this kind identified in the physical formalism, the argument goes that the human impression of fleetingly dwelling in that present moment must be an illusion. 'So much the worse for physics,' one might say in reply, 'if it proves incapable of accommodating so basic an element of the human encounter with reality.' Only someone committed to a crassly scientistic reductionism, believing that physics is all, could attempt to use such an abstract argument to discredit so basic a human experience.

One might also introduce a further scientific point into the discussion. While special relativity relates perception of time to the motion of the observer, and so declines to define a universal time that might give meaning to 'now,' when this particular universe is considered as a whole there turns out to be a natural 'frame of reference' (as the physicists say) that can be used to define a meaningful cosmic time. (The frame is defined by being at rest with respect to the universal cosmic background radiation and its existence reflects the fact that, on the largest scales, our universe is effectively homogeneous.) Cosmologists are using this definition of cosmic time when they say that the universe is 13.7 billion years old. It might seem pretty far-fetched to suppose that a cosmic concept of this kind could bear any relation to terrestrial human experience, but there is another example of what appears to be some form of linkage between the universal and the local. The insight is called 'Mach's Principle,' and it draws attention to the fact that the way matter behaves in our neighbourhood correlates with the overall distribution of matter in the universe as a whole.[3]

Proponents of a developmentally unfolding account of temporality do not only appeal to the basic human experience of the flux of time, but they also point to the causally complex and fruitfully emergent picture of physical process. I must confess myself to be of their party. A world characterised by sequential becoming does seem to be one that it is appropriate to consider as a world of intrinsic temporality, exhibiting an ontological contrast between the fixity of the past and the openness of the future.

However, a slightly tricky philosophical point also needs to be made. It is important to recognise that issues of temporality and issues of causality are, in fact, logically distinct from each other. To equate an atemporal world with determinism is to make a category mistake, for the fact is that there is no ineluctable inference to be made from atemporality to strict determinism, or vice versa. No unique pattern of causal relationships is demanded by the events of the block universe, and what those connections might actually be is a question quite separate from that of the undifferentiated existence of past, present and future. While it would have been quite natural to think atemporally about Laplace's mechanically deterministic world, in which the past and future were implicit in the present, it would not have been a forced move to do so.

Yet there do seem to be different theological implications that might be drawn from the two different pictures of the nature of time. They concern the Creator's relationship to creation. God will surely know things as they actually are, in total accordance with their nature. This seems to imply that divine knowledge of a fundamentally atemporal world would be atemporally apprehended, but a temporal world would be apprehended temporally, that is to say its events would not simply be known to be successive, but they would be known in their succession. Atemporal knowledge is precisely how classical theology thought of God's relationship to creation, believing that the whole of history is known by the divine Observer *totum simul*, all at once. All events were held to be 'simultaneously' present to the God who looks down on history from outside of time.

---

[3] Putting it more formally, local inertial frames of reference, which refer to the dynamical properties of matter in a terrestrial environment, are found to be at rest or in uniform motion with respect to the fixed stars, i.e., the distribution of matter in the universe. (That is why the period of the Earth's rotation as measured by a Foucault pendulum is the same as the length of the sidereal day.)

Temporal knowledge, on the other hand, implies a true divine engagement with unfolding time. God's creative act must then be understood to have involved the gracious divine embracing of the experience of time, the acceptance of a temporal pole within divinity. This picture seems to correspond closely to how God is portrayed in the Bible, interacting with the history of Israel and accepting a radical experience of temporality in the incarnation of the Son. This insight of divine temporality, coupled, of course, with a continuing recognition of the existence also of an unchanging eternal pole within the nature of God, has received widespread acceptance in much of twentieth-century theology.[4] It does not subvert the orthodox Christian distinction between the Creator and creation, since divine temporal polarity can be understood as a form of relationship to creatures freely accepted by God as part of the process of creation, and not simply imposed upon the divine nature.[5] The concept combines naturally with an understanding of divine knowledge as having the character of current omniscience (knowing now all that is knowable now), rather than an absolute omniscience (knowing all that will ever be knowable). This restriction would be understood theologically as being kenotic, a chosen self-limitation on the part of the Creator in bringing into being an intrinsically temporal creation. It would be no defect in the divine perfection not to know the details of the future if that future is not yet in existence and available to be known.[6]

---

[4] Process theology lays particular emphasis on divine temporal/eternal polarity, but this insight is found also in many who are outside the process fold. Process thought, however, regards this polarity as a metaphysical necessity, rather than a kenotic acceptance on the Creator's part of participation in temporality. For the scientist-theologians, see J. C. Polkinghorne, *Scientists as Theologians* (SPCK, 1996), 41. For a discussion of the approaches of Karl Barth and Eberhard Juengel to God's relationship to time, understood in the light of Christ's suffering and death and summarised in the epigram 'God's being is in becoming,' see A. E. Lewis, *Between Cross and Resurrection* (Grand Rapids, Mich.: Eerdmans, 2001), 188–95.

[5] See J. C. Polkinghorne (ed.), *The Work of Love* (SPCK/Eerdmans, 2001), 102–3.

[6] The preceding section is from *Exploring Reality* by John Polkinghorne © 2005 by Yale University, published in the USA by Yale University Press, reprinted by permission of Yale University Press, 115–19.

15

Evil

Physicists are deeply impressed by the rational order and inherent fruit-fulness of the universe.[1] Many, even among those who are not adherents to any faith tradition, incline at least to a kind of cosmic religiosity of the sort that Albert Einstein expressed when he wrote of 'a feeling of awe at the scheme that is manifested in the material universe.'[2] Hence the quite frequent, almost instinctive, recourse to the use of 'Mind of God' language when people working in fundamental physics write books for the general public.

Biologists are different. Quite commonly they display hostility towards taking any serious account of religious ideas or language. There are at least three reasons why this might be so. One is the unfortunate legacy of disputes over Darwin's evolutionary ideas, lingering even today in the circles of 'creationism' (so-called). Some deny that evolutionary understanding matters for Christian theology. In fact, quite the reverse, since respect for the truth requires Darwin's insights to be taken with appropriate seriousness. Never-theless, the memory of some of the religious mistakes of the past lingers on in the biological community, particularly among those who take no trouble to find out what contemporary theology actually has to say.

Second, placing an extraordinary degree of overconfidence in science's unaided power to gain understanding can lead some biologists to make grossly inflated claims that their insights are capable of explaining pretty well everything. Many physicists were in this kind of grandiose mood in the generations that followed Isaac Newton's great discoveries, but the later discernment of the complex subtlety of physical process eventu-ally led that community to a more humble recognition that mechanism is not all. Man is more than a machine. Yet biologists today, in the wake of their stunning discoveries in molecular genetics, are all too prone to a euphoric degree of unjustified triumphalism that grossly exaggerates the

---

[1] See J. C. Polkinghorne, *Science and Creation* (SPCK, 1988), chs 1 and 2; *Belief in God in an Age of Science* (Yale University Press, 1998), ch. 1.

[2] H. Dukas and B. Hoffmann (eds), *Albert Einstein: The Human Side* (Princeton: Princeton University Press, 1979), 70.

explanatory power of their discipline. I feel sure this is a temporary episode that will not survive a recovery of full biological interest in organisms as well as in molecules.

Yet there is also a third reason for biological reserve about religion, which is of a much more serious kind. In contrast to the austerely elegant perspective of the physicists, biologists view a scene that is much more messy and ambiguous in its character, with a mixture of fruitfulness and waste, of promise and pain. The truth-seeking explorer of reality must take this last issue with the utmost seriousness.

Of all the difficulties that hold people back from religious belief, the question of the evil and suffering in the world is surely the greatest. Narrowing the focus from nature to humanity only intensifies the issue, as the long history of war, exploitation and persecution is then brought into the perspective. How can such a world be considered to be the creation of a God who is both all-good and all-powerful? The statement of the problem is too familiar and troubling to need extensive elaboration. Not only does it give considerable pause to the enquirer after theism, but it is also one that remains a perpetual challenge and source of perplexity for those of us who are believers.

There are two different kinds of evil that need to be considered. Moral evil arises from human choices that lead to cruelty, exploitation and neglect. Natural evil arises from events outside human control, such as the incidence of disease and disaster. There is not always a clear-cut division between the two. Shoddy building practices can considerably enhance the destructive effects of earthquakes. Unjust treatment of the poor reduces their condition to an impoverished state of enhanced vulnerability to epidemics. Human lifestyle choices, such as heavy smoking, can lead to tragic early death through cancer. Yet, while the responsibility for moral evil seems to lie with human beings, ultimately the responsibility for natural evil appears to lie at the door of the Creator.

The attempt to justify the ways of God in the face of the actuality of evil is called theodicy. It is a task of considerable importance and difficulty for theologians. It is clear that the perplexities that are raised are not ones that are capable of being dispelled simply by a few paragraphs of clear-thinking prose. They are as much existential as logical and they lie very deep. Christian thought over the centuries has followed one of three basic strategies.

The first is one that the advance of science has made untenable for us today, although it was treated as very significant in the early Christian centuries. A plainly literal reading of Genesis 3.14–19 (the words of God to Adam and Eve and the serpent in the mythic story of the eating of the forbidden fruit and its aftermath) led to the idea that the Fall, understood as the original act of moral evil, also resulted in a curse upon creation that was the actual source of natural evil. Paul appears to write within this kind of understanding when he speaks of Adam as the one through whom sin came into the world 'and death came through sin' (Romans 5.12). It is obvious that our knowledge of the long history of life, with the mass extinctions that have punctuated it, does not permit us today to believe that the origin of physical death and destruction is linked directly to human disobedience to God. However, if we understand the story of the Fall to be the symbol of a turning away from God into the self that occurred with the dawning of hominid self-consciousness, so that thereby humanity became curved-in upon itself, asserting autonomy and refusing to acknowledge heteronomous dependence, we can today interpret those words in Romans in the sense of referring not to fleshly death but to what may be called 'mortality,' spiritual sadness at the transience of human life.[3] Because of their self-conscious power to look ahead into the future, our ancestors had become aware that they would die. This was an emergent recognition of something always present, namely the finitude of life in this world.

Christian belief embraces the idea that God's purposes will find their ultimate fulfilment beyond present history in the everlasting life of the world to come, but the Fall meant that our ancestors had become alienated from the One who is the only true ground of hope for that *post mortem* destiny. Hence their feeling of the bitterness of mortality, an experience in which we also share, for we are the heirs of that fractured relationship with our Creator. This modern interpretation of the Fall and its consequences conveys an important insight into the human condition, but it does not, in itself, offer us a resolution of the problem of evil.

The second strategy of theodicy is an attempt to deny the absolute reality of evil. It is claimed that evil is no more than a kind of deprivation, the absence of the good rather than the substantial presence of the bad – rather as darkness is simply the absence of light. (There are photons, particles of light, but there are no scotons, particles of darkness.) After the terrible events of the twentieth century this seems to me to be an impossible stance to adopt. In fact, when one considers an appalling episode like the Holocaust, though one can see individual and societal

---

[3] J. C. Polkinghorne, *Reason and Reality* (SPCK/Trinity Press International, 1991), ch. 8; *Belief in God in an Age of Science*, 88–9.

factors at work (the implacably evil will of powerful leaders; a society in which an unquestioning obedience to the State had been strenuously inculcated; ordinary human cowardice that meant that people looked the other way when the cattle trucks laden with their human cargo rumbled through the village railway station on the way to Auschwitz), nevertheless there is a weight of evil involved in these dreadful events that makes me, at any rate, not quick to be dismissive of the possibility that there are also non-human powers of evil loose in the world.

If that is so, it does nothing of itself to resolve the problems of theodicy, since the question of how these satanic powers originated, and why they are permitted to continue, remains deeply troubling. Whatever view one takes about the nature of spiritual evil, it seems that evil's reality is just too great to be argued away as simply the privation of the good. Yet, having acknowledged that, the light/dark comparison does serve to remind us of the existence of very much positive good in the world, so that the problem of evil has to be held in tension with the 'problem' of the existence of value and good. The world is both beautiful and ugly, inspiring and terrifying in turn.

The third strategy of theodicy is the one followed by most contemporary theologians. It seeks to make out a case that the evils that occur are the necessary cost of greater goods that could be attained in no other way and which more than redress the balance of creation in God's favour. According to this view, the dark side of creation is the unavoidable shadow that is inseparable from its goodness. In relation to moral evil, this argument is summed up in the well-known free-will defence: a world with freely choosing beings, however bad some of their choices may prove to be, is a better world than one populated only by perfectly programmed automata. This is not a claim that can be made in this post-Holocaust era without a quiver in the voice. Nevertheless, I believe that there is important truth here. We instinctively recognise that acts that seek to manipulate and restrain a person's freedom of action, even when undertaken with desirable intentions, such as various acts of restraint laid upon potential or actual offenders in order to avoid permitting the infliction of harm, are in themselves acts of imperfection, in that they diminish the humanity of those on whom they are imposed. Philosophers argue whether or not it would have been possible for God to have created beings who *freely* and *always* choose the good. There does seem to be a paradox in this notion.[4]

---

[4] The preceding section is from *Exploring Reality* by John Polkinghorne © 2005 by Yale University, published in the USA by Yale University Press, reprinted by permission of Yale University Press, 136–42.

* * *

A somewhat similar appeal to the necessity of a distance between the Creator and creation can be made in relation to the problem of natural evil. Rather than this world being a ready-made divine puppet theatre, we have seen that its character of being the home of an evolving process can be understood theologically as showing it to be a creation in which creatures are allowed 'to make themselves.' This seems indeed to be a great good, but it also has a necessary cost. As the generations succeed each other in the course of evolutionary process, death is seen to be the prerequisite of the possibility of new life. The history of the shuffling exploration of potentiality will inevitably have its ragged edges, for there will be developmental blind alleys and extinctions, as well as unfolding fertility.

Another way of putting the point is to frame what I have called 'the free-process defence:'[5] all of created nature is allowed to be itself according to its kind, just as human beings are allowed to be according to our kind. As a part of such a world, viruses will be able to evolve and cause new diseases; genes will mutate and cause cancer and malformation through a process that is also the source of new forms of life; tectonic plates will slip and cause earthquakes. Things will often just *happen*, as a matter of fact, rather than for an individually identifiable purpose. The question so often asked of a minister by those who are in great trouble, 'Why is this happening to me?' may sometimes have no answer beyond the brute fact of occurrence.

Science can offer some help to theology here in support of the necessary cost of a world allowed to make itself. We tend to think that had we been in charge of creation, frankly, we would have done it better. We would have kept all the nice things (fruitfulness and beauty) and got rid of all the nasty things (disease and disaster). However, the more science enables us to understand the nature of evolving fertility, the more we see that it is necessarily a package deal, an integrated process in which growth and decay are inextricably interwoven as novelty emerges at the edge of chaos. The ambiguous character of genetic mutation, both the engine of evolutionary fruitfulness and the source of malignancy, illustrates the point.

A theologian would say that what is involved in the occurring costliness of creation is the divine permissive will, allowing creatures to behave in accordance with their natures. Bringing the world into being was a

---

[5] J. C. Polkinghorne, *Science and Providence* (SPCK, 1989), 66–7.

kenotic act of self-limitation on the Creator's part, so that not all that happens does under tight divine control. The gift of Love in allowing the genuinely other to be is necessarily a precarious gift. I believe that God wills neither the act of a murderer nor the incidence of an earthquake, but both are allowed to happen in a creation given its creaturely freedom.

There may seem to be something very bleak in such a conclusion, but I think that it represents the necessary primal reality of a world not yet fully integrated with the life of God. The free-will and free-process defences are just two sides of one coin, the cost of a world given independence through the loving gift of its Creator. The two insights are also linked by the fact that the possibility of the morally responsible exercise of free will depends upon its taking place in a world of sufficiently stable integrity that actions can have foreseeable consequences. The ethical imperative of care for others would become meaningless if God could always be relied upon for magical interventions to save people from the bad consequences of human carelessness and neglect.[6]

* * *

Part of the problem of evil is simply its scale. Some degree of danger and struggle could be seen as providing a challenging spur to growth and development, but too often suffering seems only to diminish or extinguish the humanity of those on whom it falls. There is a mystery here that will not yield simply to rational analysis. In reality, the problem of evil is too profound to be dealt with adequately by any form of moral bookkeeping, as if one were simply casting up creation's ethical profit-and-loss account. Much of the discourse of philosophers on this issue, whether of theistic or atheistic stripe, is too coolly detached to carry much conviction.[7] The precise quantification of evil is a highly problematic notion, even if one can see that there are greater and lesser ills.

Ultimately, responding to the surd of tragedy requires the insights of the poet more than the arguments of the logician. Important for me is the passion of Christ, understood as divine participation in the travail of creation. Here is a point unique to Christianity, with its trinitarian and incarnational understanding of the nature of God. One might dare to say

---

[6] The preceding section is from *Exploring Reality* by John Polkinghorne © 2005 by Yale University, published in the USA by Yale University Press, reprinted by permission of Yale University Press, 143–5.

[7] See, for example, E. Stump and M. J. Murray (eds), *Philosophy of Religion: The Big Questions* (Oxford: Blackwell, 1999), 151–262.

that the burden of existential anguish at the suffering of the world is not borne by creatures alone, but their Creator shares the load, thereby enabling its ultimate redemption. Christianity is a religion that often calls for the acceptance of suffering, in contrast to the Buddhist counsel to flee suffering, and it does so because it can speak of that acceptance as a participation in the sufferings of Christ (1 Peter 4.12–19). The Christian God is the crucified God, not a compassionate spectator from the outside but truly a fellow sufferer who understands creatures' pain from the inside. Only at this most profound level can theology begin truly to engage with the problem of the evil and suffering of this world.[8]

---

[8] The preceding section is from *Exploring Reality* by John Polkinghorne © 2005 by Yale University, published in the USA by Yale University Press, reprinted by permission of Yale University Press, 145–46.

Part 3

CHRISTIANITY

16

Scripture

Those of us who write about the traffic across the border between science and religion are sometimes reprimanded by reviewers, usually of an evangelical persuasion, for paying insufficient attention to the Bible. It is not being suggested that the answers to modern scientific questions are to be found within its covers – for the curious and disturbing phenomenon of 'scientific creationism' is scarcely to be encountered in Britain, however much it may rampage on the other side of the Atlantic – but rather that questions of reason and reality should be judged from a standpoint controlled by Scripture. Here, it is proposed, lies the key to theological interpretative problems. Stated bluntly like that, the proposal seems to short-circuit an obvious difficulty. I gladly acknowledge the importance of Scripture for Christian life and thought, but how is it actually to be used? The briefest acquaintance with the history of the Church makes it plain that a variety of methods have been employed, from the narrowest literalism to the most fantastic allegorization, from reliance upon isolated proof texts to the most generalized notion of a record of evolving religious consciousness. As part of our exercise of reason in the pursuit of reality we have to consider what is the proper usage of the Christian Scriptures.

At the outset, let me say that one of the reasons why detailed attention to the Bible may play only a subsidiary role in much writing about science and theology is that such writing itself plays only an auxiliary role in relation to the great endeavour of the intellectual exploration of Christian faith. I do not say that such writing is not of importance – for me it is an essential task to hold together my scientific and Christian insights with as much integrity as I can muster, and to some extent that must be true for those who are not themselves scientists but who live in a culture in which science is a significant component. I do say, however, that considerations of natural theology and the like do not afford the fundamental basis for my own religious belief. That lies in my encounter with God in Christ, mediated through the Church, the sacraments, and, of course, the reading of Scripture. The discussion of science-and-religion is a valuable but second-order task, in which one seeks an harmonious integration of one's basic experiences as a believer and as a scientist. In a sense it is a fringe activity, selecting from each area those elements which lie closest to the other, and not necessarily reflecting more than a part of that subject's central preoccupations. One no more expects to get from such writing a balanced account of theology than one supposes its discussion of science to represent an evenhanded survey of the physical world. In each case, the material selected is chosen with the other discipline in mind.

I have already acknowledged that the Bible is important to me in my own religious life. There is a strong tradition of Christian thought which assigns it a supreme role, making the Bible the arbiter of all theological inquiry. A conservative biblicism has often proved attractive to scientists, particularly in their student days. They are familiar with the notion of the textbook, that reliable source of information in which one can look up the answer to one's queries. Much painful labour can be avoided in that way, and there is a certain attraction in the feeling that God should have provided just such a textbook to help us with our religious search. Those who go on to postgraduate scientific study learn that even in the physical world our explorations do not always lead us to cut-and-dried answers. We are not on our own, for we benefit from the accumulated experience and discovery of the scientific community, and its record in the libraries, but even the most assiduous perusal of the pages of the *Physical Review* will not tell us all that we need to know. The search for truth involves more than the ability to scan the literature.

It is important to recognize that, though almost all Christians down the centuries have assigned a significant and normative role to Scripture, they have not been 'people of the book' in the way that our Islamic friends are, who regard the Koran as a divinely dictated document (and so only properly to be read in the original classical Arabic), or even in the way of Rabbinic Judaism, with its appeal to the Torah (however overlayed with Talmudic interpretation, understood as deriving from a parallel oral tradition). The classic Christian attitude to the Bible has been a subtle mixture of respect and freedom, and it was so from the beginning. Jesus contrasts the creation ordinance of marriage as the union of husband and wife in one flesh with the Mosaic ordinance permitting divorce, and adjudicates in favour of the former (Mark 10.2–9). In the Sermon on the Mount he quotes what 'was said to the men of old' (in fact, in the Torah) and deepens or radicalizes it ('But I say to you . . .') (Matthew 5).[1]

---

[1] The preceding section is from *Reason and Reality* by John Polkinghorne © 1991 by John Polkinghorne, published in the USA by Trinity Press International, 1991, 60–1.

* * *

In our day, the New Testament is one of the great vehicles which bring about our encounter with Christ. For the Christian the true Word of God is written, not with paper and ink, but in the flesh and blood of that life lived in Palestine long ago (John 1.14) and in the continuing life of the Risen Lord. All authority rests with him (Matthew 28.18) and it is not located between the covers of any book. Yet it is from the Bible that we shall learn much of Jesus and of the life of Israel into which he entered. 'We can account adequately for the position that Scripture holds in the Christian faith,' says Barton, 'only on the basis of a belief that God was genuinely and uniquely known in Israel and was then made more profoundly known in Jesus Christ.'[2] Because revelation is the encounter with a Person and not the deliverance of a set of propositions, the Bible is not our divinely-guaranteed textbook but a prime means by which we come to know God's dealings with humankind and particularly his self-utterance in Christ.

A number of things then follow. The first is that there will be an evidential role for Scripture. It is necessary for us to have grounds for the belief that God was in Christ, and it is important for us to know what Jesus was like. 'The Bible matters, to put it at its simplest,' says Barton, 'because it is the earliest and most compelling evidence that Jesus rose from the dead, and that he was such a person that his rising from the dead is gospel, good news.'[3] As James Barr has said, 'Few would be willing to rest content with a Jesus who in historical fact was an unprincipled crook, a used chariot salesman of the time.'[4]

Many theologians recoil from acknowledging this historico-evidential role for Scripture. They would prefer its being regarded as kerygmatic, the proclamation of a saving faith to be accepted as transcendentally given (Barth) or to be embraced existentially (Bultmann). Otherwise they fear that what should be timeless truth is hazarded upon the uncertain judgement of historians as to what actually happened. But the religion of the incarnation is based on the idea that the Eternal made himself known in the contingency of the temporal, taking the risk of an involvement with human history (Philippians 2.6–8). However moving the story, *simply as story* it is not enough. We need to know it happened. It is instinctive for the scientist to ask the question 'What is the evidence? What makes you think it is actually the case?'

---

[2] J. Barton, *People of the Book* (Atlanta: Westminster John Knox, 1988), 21.
[3] Ibid., 40.
[4] Quoted in Barton, *People of the Book*, 42.

The hermeneutics of suspicion has presented our generation with a temptation to take refuge from history in evoking the power of language, to rest content with the 'language event.' 'The classical mind says that's only a story,' says Crossan, 'but the modern mind says, there's only story.'[5] I am not willing to resign so easily from the cognitive quest. I cannot accept the view, described by Northrop Frye, that 'the events the Bible describes are what some scholars call "language events," brought to us only through words; and it is the words themselves that have the authority, not the events they describe.'[6]

I do not, of course, deny the presence of story in the Bible (Jonah, Daniel, and so on). But the life, death and resurrection of Jesus is not just a tale, however evocative, but a wonderful fusion of the power of myth and the power of actuality. Nor do I deny the problems presented by the task of historical assessment. The endeavour calls for scrupulosity and delicacy of judgement, though I think that many modern scholars have greatly exaggerated the degree of difficulty involved to the point of claiming it to be virtually impossible. This is not the place to attempt to present my own conclusions from looking into the matter,[7] but I simply record that I believe, in particular, that the New Testament can be used as an evidential basis for supporting the claims of the Christian faith about its Founder, Jesus Christ.[8]

The Old Testament presents us with greater problems. Yet, not only is the absorbing account of intrigue at the court of King David (2 Samuel 11—1 Kings 2) an extremely early example of detailed historical writing (probably antedating Herodotus by at least four centuries), but I believe that behind the multi-layered story of the foundational experience of the Exodus lies the historical reminiscence of a great act of deliverance.

Of course, the bare events themselves are not enough. Even the empty tomb (Matthew 28.5–7; Mark 16.6; Luke 24.5–7; John 20.4–9) and the resurrection itself (Matthew 28.16–20; Luke 24.38–49; John 20.19–23) need interpretation. As in science, so in theology, experience and interpretation intertwine.[9]

5 J. D. Crossan, quoted in J. M. Soskice, *Metaphor and Religious Language* (Oxford: Clarendon Press, 1985), 109.

6 N. Frye, *The Great Code* (London: Routledge & Kegan Paul, 1982), 60.

7 J. C. Polkinghorne, *The Way the World Is* (Eerdmans, 1984), ch. 6.

8 See C. H. Dodd, *The Parables of the Kingdom* (Welwyn: James Nisbet, 1961).

9 The preceding section is from *Reason and Reality* by John Polkinghorne © 1991 by John Polkinghorne, published in the USA by Trinity Press International, 1991, 62–4.

* * *

It seems clear to me, as a matter of experienced fact, that the Bible, despite all its cultural strangeness and scientific inadequacy, does actually succeed in speaking to us across the centuries. Moreover, this engagement with another thought world can be liberating, in that it points us away from the limitations of a merely twentieth-century world-view. It would be an act of intellectual arrogance to suppose that our insight is in every respect superior to that of preceding generations. They may well have known things now forgotten or grown dim, not least in the realm of spiritual experience.

Just as with our contemporaries there is 'the perfectly rational enterprise of using the wider resources of the community to extend one's own, and necessarily limited, experience and expertise,'[10] so we can carry that procedure further still in our dialogue with those who are our predecessors. Such contact is, in any case, essential in order that we may recognize the womb that gave us birth. Christians are not called to practise a false antiquarianism, as if belief required an anachronistic attempt to become a first-century Palestinian, but they have to be in touch with the foundations of their faith, 'the apostles and prophets, Christ Jesus himself being the cornerstone' (Ephesians 2.20). We must be in contact with our origin. Here is a great contrast with science. I do not need to read Clerk Maxwell's classic work, the *Treatise on Electricity and Magnetism*, in order to make use of his equations. I do need to read the Gospels if I am to reckon with Christ.[11]

* * *

The ability of literature to speak with meaning across the centuries is a phenomenon familiar to us and by no means confined to Scripture. We encounter it in Shakespeare and the Greek tragedians. It is one of the principal arguments for a continuity of human nature lying beneath the bewildering varieties of cultural change. It is expressed in the idea of the classic, which Schlegel spoke of as 'a writing which is never fully understood. But those that are educated and educate themselves must

---

[10] Soskice, *Metaphor and Religious Language*, 152.

[11] The preceding section is from *Reason and Reality* by John Polkinghorne © 1991 by John Polkinghorne, published in the USA by Trinity Press International, 1991, 65.

always want to learn from it.'[12] There is a close connection between this inexhaustible character of the classic and the unlimited open fruitfulness of symbol. The universality of the classic means that it is never the possession of a sect, for it cannot be confined within narrow limits. David Tracy speaks of the classic as 'always public, never private,' and he goes on to say that in the classic 'we recognize nothing less than the disclosure of a reality we cannot but name truth.'[13] The classic maintains an equipoise between then and now; it is fruitful of novelty without denying its origin in the past. As John Barton says, 'it remains itself, yet has something fresh to say to each new inquirer.'[14]

When we think of the Bible as being the supreme Christian classic, we are encouraged to read it in a particular way. I have spoken of its evidential role. In that historical mode, our reading must be critical and analytical; a concern with questions of sources and authenticity is paramount. The writing is being submitted to our evaluation. When we read Scripture in a classic mode we are submitting ourselves to it. Those of us who use the Bible in our spiritual life are engaged with the text in a way that allows it to become part of us, forming the fabric of our thought. We are experiencing its power to be the vehicle used by the Spirit to bring God's grace to us. Our concern is not then analytical, dividing the text into fragments of differing kinds and qualities, but with the totality of what is set before us. David Tracy says, 'explicitly religious classic expressions will involve a claim to truth by the *power of the whole* – as in some sense a radical and finally gracious mystery.'[15] That will be true of the literary corpus of the Bible as well as of the world-view of Christian faith. The recognition that this is so is one of the central tenets of recent 'canonical criticism'[16] encouraging us to take seriously the whole and final form of Scripture.

Of course the Old Testament is a combination, often within the same book, of material from different centuries, so that Yahweh is sometimes spoken of parochially as Israel's God and sometimes universally as the God of all the Earth; of course there are distressing harshnesses in tales of genocide and stoning to death, and in the vindictiveness of the cursing psalms, which mean that they cannot be taken as Christian exemplars; of

---

[12] Quoted in B. Lonergan, *Method and Theology* (London: Darton, Longman & Todd, 1972), 161.

[13] D. Tracy, *The Analogical Imagination* (London: SCM Press, 1981), 7 and 13.

[14] Barton, *People of the Book*, 63.

[15] Tracy, *The Analogical Imagination*.

[16] Brevard S. Childs, *Old Testament Theology in a Canonical Context* (Philadelphia: Augsburg Fortress, 1985), 13.

course there are contrasting points of view in the New Testament (compare the attitudes of Romans 13 and Revelation 13 to the governing authorities or, more centrally, the differing approaches to Christology found in John, Paul and the Writer to the Hebrews);[17] but the bewildering richness and conflict of life is present in this sometimes dissonant diversity making up the Bible, and beneath it all, as its ground base and unifying principle, is continuing testimony to the steadfast love of God and his Christ.

The presence of clash and contradiction within the Bible is fatal to the theory of using it as a divinely-guaranteed textbook. Those who attempt to do so either construct a tacit canon within the canon, omitting what is awkward and unseemly, or they are driven to frustration or crazy ingenuity in trying to reconcile the irreconcilable, to square the God of love with the command to Saul utterly to destroy the Amalekites (1 Samuel 15). If the reading of Scripture is to be truly edifying it will have to be in some other mode than this. The clue lies, I believe, in the assimilation of our engagement with the Bible to our fundamental experience of open engagement with symbol. 'The Biblical text mediates not information or opinion but encounter,' says John Barton.[18] Symbol is not to be reduced to sign by an insistence that it carry a single univocal meaning.

Equally the Bible is not to be tied down; it must be acknowledged as being polysemous, having multi-layered meaning, capable of mediating many messages to its readers. Barton tells us that 'the semantic indeterminacy of sacred texts'[19] is a continuing theme of his book, which sets out to discuss the nature of 'the authority of the Bible in Christianity.' Time and again we see this verbal fruitfulness blossoming within Scripture itself. Flattering (or reassuring) words are spoken at the coronation of a king of Judah: 'The Lord has sworn and has not changed his mind, "You are a priest for ever after the order of Melchizedek"' (Psalm 110.4). In the mind of the Writer to the Hebrews these words trigger the sustained meditation on the eternal and effective priesthood of Christ, which is the central theme of his epistle. Melchizedek himself, a shadowy but impressive figure from patriarchal history (Genesis 14), becomes the type of Christ, 'the source of eternal salvation to all who obey him' (Hebrews 5.9). A contemporary commemoration of some unknown righteous sufferer in the period of the Exile, or conceivably of the exiled community itself (Isaiah 53), provides the first Christians with the key to understanding the paradox of a crucified Messiah.

---

[17] J. D. G. Dunn, *Unity and Diversity in the New Testament* (London: SCM Press, 1977).
[18] Barton, *People of the Book*, 57.
[19] Ibid., 20.

The author's particular original intention is not the only meaning to be found in the words used, and the etymological fallacy (that original sense is the true sense) is no more to be imposed on passages than it is on individual words. In the biblical examples cited, there is an appropriateness of development by which intuitions of priesthood, and of the redemptive character of vicarious suffering, find their focus and fulfillment in Christ. 'Types and shadows have their ending.'

This process of conceptual development does not stop with the end of the New Testament era. The Fathers recognized the varieties of meaning to be found in Scripture – the literal, the moral, and the spiritual, the last two often uncovered using methods either allegorical or typological – even if some of their exegesis may seem to us to be either mechanical or fantastic. Polysemy is an indispensable element in the inexhaustible freshness of the religious classic, the Bible. Northrop Frye says of polysemous meaning, that as we read 'the feeling is approximately "there is more to be got out of this".'[20] The process is not one of willful imposition but of sympathetic exploration. Frye emphasizes continuity of development. 'What is implied is a single process growing in subtlety and comprehensiveness, not different senses, but different intensities of a continuous sense, unfolding like a plant out of a seed.'[21] It is this naturalness that we often feel to be missing in the allegorical extravagances of interpreters like Origen. The overplus of meaning is surely to be sought without undue strain. Yet, despite the inevitable oddities of individual expositors, the evolving tradition of the Church provides a context of the kind of continuously developing exploration to which Frye refers. It is the setting within which those of us who believe in the guidance of the Spirit will wish to pursue our own use of Scripture, not in slavish deference to the past but in the recognition that we need to profit from its insights.[22]

* * *

What has the Bible to say to those who seek to use it in this way in order to cast light on the relationship between science and theology? The opening passages of three of the writings of the New Testament set before us the image of the Cosmic Christ: the Word (much more than rational

---

[20] Frye, *The Great Code*, 220.
[21] Ibid., 221.
[22] The preceding section is from *Reason and Reality* by John Polkinghorne © 1991 by John Polkinghorne, published in the USA by Trinity Press International, 1991, 66–8.

principle but surely including the idea of such a principle) without whom 'was not anything made that was made' (John 1.3); the One of whom it can be said 'all things were created through him and for him. He is before all things and in him all things hold together' (Colossians 1.16–17); the One 'through whom also [God] created the world' (Hebrews 1.2). What greater encouragement could there be for the scientific exploration of the rational structure of the physical world, what clearer indication of its value?

In an astonishing passage in Romans, Paul extends to all creation the hope of an eventual liberty in Christ:

> For the creation waits with eager longing for the revelation of the sons of God; for the creation was subjected to futility, not of its own will but by the will of him who subjected it in hope; because the creation itself will be set free from its bondage to decay and obtain the glorious liberty of the children of God. We know that the whole creation has been groaning in travail together until now; and not only the creation, but we ourselves, who have the first fruits of the Spirit, groan inwardly as we wait for adoption as sons, the redemption of our bodies. (Romans 8.19–23)

Here is set before us no narrow anthropocentric hope but one of literally universal proportions. Earlier Paul had affirmed that something can be known of God from the inspection of the world 'namely his eternal power and deity' (Romans 1.19), a significant encouragement to pursue the insights of natural theology.

The New Testament, written by city dwellers, has comparatively little to say about nature itself, though Jesus uses natural imagery in sayings and parables. It is the Old Testament which provides many detailed appeals to the world we inhabit. They are particularly to be found in the Wisdom literature – that cool look at the way things are, which I have suggested is the form that natural theology takes in the Hebrew Scriptures and which is the nearest that Israel got to anything remotely approaching science.[23] Most striking of all such passages is the answer to Job (Job 38—41). The voice from the whirlwind, addressing that righteous sufferer, speaks mainly of an appeal to the power and mystery and 'otherness' of created nature. 'Where were you when I laid the foundations of the earth? Tell me if you have understanding' (Job 38.4). Once again there is a corrective to narrow anthropocentrism; a strong sense of God's varied purposes for all his creation pervades the discourse. The psalms frequently celebrate God's

---

[23] J. C. Polkinghorne, *Science and Creation* (SPCK, 1988), 3–6.

power as Creator, and the fruitfulness of his creation (Psalms 8, 19, 29, 33, 65, 96, 104, 135, 136, 139, 147), not in a sentimental way but in one which recognizes the destructiveness of the stormwind and that 'the young lions roar for their prey, seeking their food from God' (Psalm 104.21). The Hebrew mind is always aware of the threat posed by the waters of chaos, held in restraint by God alone (e.g. Psalm 93.3–4). Anyone constructing a service on a scientifically oriented theme will find much more choice of material in the psalter than in the hymnbook.

The dominant image is the value of creation and the power of its Creator, providing further encouragement to scientific exploration. A similar theme recurs in the prophetic writing most concerned with these matters, Second Isaiah. God is the ground of all that is: 'For thus says the Lord who created the heavens (he is God!), who formed the earth and made it (he established it; he did not create a chaos, he formed it to be inhabited!), "I am the Lord, and there is no other"' (Isaiah 45.18). He is no deistic God content with what he has done, but he is active, his care continues: 'Behold, the former things have come to pass, and new things I declare; before they spring forth I tell you of them' (Isaiah 42.9). God's world is one of dynamic becoming, not static being (as science has indeed discovered it to be). The comparison with the endlessly cyclic nature-religions of Canaan and Babylon, and the perpetual status quo of Egypt, is very striking.

I have left until last what many believe to be the *locus classicus* of the interaction of science and Scripture, the opening chapters of Genesis. Seldom does one give a talk about science-and-religion to a general audience without being asked how these chapters come into it. Nothing indicates more clearly than they do the need to use Scripture aright. Part of our respect for the Bible must involve our attempting to read the different parts of it in the appropriate fashion. It is a disastrous mistake to read poetry as if it were prose. 'My love is like a red, red rose' does not mean that I am in love with Ena Harkness!

Nowhere is the textbook approach to Scripture more out of place than in Genesis 1—3. We are alerted to this fact by noting that two different creation stories have been juxtaposed to each other, an early account (Genesis 2.4–24) and a later one of considerable sophistication[24] (Genesis 1.1—2.3). When Genesis reached its canonical form, the redactor did not feel it necessary to conflate or reconcile these stories, as if they had

---

[24] One is impressed by the absence of such embarrassing detail as one finds in the *Enuma Elish* (the Babylonian myth with which Genesis is often compared), where Marduk slices the defeated Tiamath in two, forming earth and sky of the two halves.

been literal accounts which must be squared with each other. That in itself tells us something of how they are to be used. We read them as powerful symbolic stories (myths) conveying the idea of a total dependence of the creation upon its Creator and (most astonishing of all) the sevenfold reiterated message that all is 'good.' Science, in making untenable a literal reading of Genesis 1 and 2 (itself a tendency originating in late medieval and reformation times), has liberated these chapters to play their proper and powerful role in Christian thought.

In a similar way, Genesis 3 is to be understood as a myth about human alienation from God, and not the aetiological explanation of the only too evident plight of humanity. Science can discern no radical rift in the course of cosmic history, brought about by primeval human error. Theology can discern the truth that the root of sin is the refusal to accept the status of a creature, to seek on the contrary 'to be like God' (Genesis 3.5). Jesus is the second Adam (Romans 5.12–17; 1 Corinthians 15.22), not as the reversal of some genealogical entanglement, but because his death and resurrection restore to us the way back to the God from whom we have wandered, and they are the initiating events of God's new creation. I believe that the correct use of Scripture not only delivers us from the tortuous trickery of 'creation science' but also encourages us to find in the opening chapters of the Bible a profound source of spiritual challenge and illumination.

Exegetes differ about whether Genesis 1.1–3 presents a picture of God creating order out of primeval chaos (the *tohu wabohu*, 'without form and void,' of verse 2), or whether it can be seen as pointing to the later Christian doctrine of *creatio ex nihilo*. The earliest unequivocal statement of the idea of creation out of nothing is in the apocryphal Second Book of Maccabees (2 Maccabees 7.28), but the emphasis in Genesis 1 on the dependence of all upon the sovereign will of God for its existence ('And God said "Let there be . . ."') is certainly consonant with the central significance of *creatio ex nihilo*. Many modern theological writers wish to combine that emphasis on creaturely ontological dependence with a notion of *creatio continua*, God's unfolding purposive action through the evolving history of the universe.[25] That sits somewhat uneasily with the 'seventh day' of sabbath rest (Genesis 2.1–3), with its implication of the completed work of creation, though some ingenious commentators suggest that cosmic sabbath is yet to come. Continuing creation is, of course, perfectly consonant with the talk of God's new things which we find in Second Isaiah. A discriminating

---

[25] See T. Peters (ed.), *Cosmos and Creation* (Nashville: Abingdon Press, 1989).

use of biblical imagery is surely legitimate for us as we seek to integrate what we learn from the book of Scripture with what we learn from the book of nature. Such discrimination is part of the proper use of Scripture.[26]

---

[26] The preceding section is from *Reason and Reality* by John Polkinghorne © 1991 by John Polkinghorne, published in the USA by Trinity Press International, 1991, 70–3.

17

The historical Jesus

The books of the New Testament in general, and the gospels in particular, must be the most researched and continually reassessed writings ever submitted to scholarly scrutiny. For more than two hundred years, a great industry of learning has busied itself with that critical analysis. I like to read as much as I can of that kind of writing but, of course, I can claim no expert status in speaking on these issues. How then can I have the temerity to address the questions? One answer would be that I do not have any choice. A Gifford Lecturer must be a bottom-up thinker. He must start with the phenomena – and the foundational phenomena of Christianity are set out in the New Testament. But it is not mere necessity which drives me to the task. While I respect and value the insights that scholars provide, I cannot believe the matter should be left solely in their hands. If we were presented with a substantial body of 'agreed results,' that submission to scholarly authority might be possible. But the fact of the matter is that we are not. A survey of thought about the New Testament[1] soon reveals the clash of view and the ebb and flow of fashion so characteristic of any scholarly activity in which, through successive generations working with limited primary resources, each of those generations seeks to establish its own originality. I do not for a moment deny the truthful intent of this labour, or that each generation succeeds in some way in deepening our understanding. But I am conscious of the effect of the pressing academic need to say something new (so familiar in my own field, theoretical physics) and also of the special problems in this area resulting from our inability to agree on widely accepted criteria by which results could be validated.

This lack of agreement derives, I believe, partly from the significance of the issues at stake. Historical judgements inevitably call for a high degree of the use of those tacit skills whose exercise we have already recognized as being an essential part of the rational pursuit of knowledge.

---

[1] See, for example, S. C. Neill, *The Interpretation of the New Testament* (Oxford: Oxford University Press, 1964).

It is not surprising, therefore, that there should be difference of opinion about which gospel sayings are authentically to be attributed to Jesus. What is surprising is the degree of such disagreement. Ancient historians have sometimes criticized New Testament scholars for their marked reluctance to trust their sources.[2] But in many people's minds, much more hangs on the assessment of the life and character of Jesus than hangs on the assessment of the life and character of Julius Caesar. What we think about him *matters*. One result of this is to demand in the former case a greater degree of scrupulosity than is required in the latter case; another result is that in the case of Jesus a hidden agenda of wider significance operates to influence the mode of ostensibly historical judgement. This hidden agenda can generate either an undue skepticism or an undue literal reliance.

In the end, I do not think one can divorce New Testament scholarship from Christian theology, any more than one can divorce experiment from theory in science. The current academic habit of respectfully observing such a division in the theological world has not been helpful. We need more people bold enough to venture across scholarly boundaries. I think that extends even to amateurs like myself, who have certain rights to talk about the general character of the wood, even if they do not have the expertise correctly to identify every one of its individual trees. Without further apology, I direct myself to the task.

The gospels are an idiosyncratic sort of writing. Their concern is with Jesus of Nazareth, but they are clearly not biographies in a modern manner. Not only do they fail to record all sorts of things that we, with our modern concerns, would rather like to know (what did he look like?), but they are clearly written from a point of view for a particular purpose. The very word *evangelion*, asserted in the first verse of the earliest gospel[3] (Mark 1.1), means 'good news' and, for his part, the author of the fourth gospel is clear what he is about: 'these are written that you may believe that Jesus is the Christ, the Son of God, and that believing you may have life in his name' (John 20.31). Yet the form of this good news is the story of a life and death and its aftermath. There is a clear intent to root the gospel in the events of history.[4]

---

[2] A. N. Sherwin-White, *Roman Law, and Roman Society: The Sarum Lectures 1960–1* (Grand Rapids, Mich.: Baker; Oxford: Oxford University Press, 1963).

[3] I follow conventional thinking about questions dating New Testament documents; but see J. A. T. Robinson, *Redating the New Testament* (London: SCM Press, 1976) for different views.

[4] Cf. N. T. Wright, *The New Testament and the People of God* (London: SPCK, 1992), especially 396–403.

We are familiar with the distinction in character between the three synoptic gospels and the fourth gospel: the synoptic Jesus, uttering pithy sayings and telling pungent parables, a first-century Palestinian proclaiming the kingdom of God; the Johannine Jesus, speaking in timeless tones, whose long discourses concern, not the kingdom, but himself, proclaimed as the Good Shepherd and the True Vine. Doubtless there is a mingling in John of the historic Jesus and the post-resurrection exalted Christ, and perhaps that admixture is not wholly absent from the other gospels either, but we must remember that John chose to write a *gospel* (that is, an account centered on a life and not just a series of timeless meditations), and that in geography and chronology we have good reasons for taking his version very seriously.[5] History is closely woven into the fourth gospel (e.g., archaeological research has shown that the five porticoes of the pool by the sheep gate (John 5.2) were actually there and they are not a symbol for the five books of the *Torah*, as some commentators had fancifully supposed).

There is neither a pedestrian literalism nor a cavalier disregard in the gospels' attitude to historical events. It seems to me that the writers are neither slavish about detailed accuracy nor careless about what actually happened. As to the first point, it is instructive to compare two versions which most of us would think originate from the hand of the same author. Luke, in his gospel, claims to write an 'orderly account' (Luke 1.3), and in Acts, by implication, to continue it (Acts 1.1), but it does not seem to trouble him that in the first book Jesus parts from his disciples on the first Easter day (Luke 24.51; taking the longer reading of the text) while in the second book this parting is dated forty days later (Acts 1.3). Of course, the ascension is a peculiar kind of event in which the symbolic predominates over the literal, but one sees a rather similar looseness of chronological attribution in the well-known conflicts between the synoptics and John as to the dating of the cleansing of the Temple and the relations of the crucifixion to the day of Passover.[6]

On the other hand, there are many indications in the gospels of a respect which the authors had for elements of the historic tradition, even when they were puzzling or embarrassing from the perspective of those writing a generation or two after the events themselves. Mark records Jesus as replying to a man who addressed him as 'Good Teacher,' by saying, 'Why do you call me good? No one is good but God alone' (Mark 10.18). Not a remark you would expect to be perpetuated about a revered leader, unless

---

[5] C. H. Dodd, *The Parables of the Kingdom* (Welwyn: James Nisbet, 1963).

[6] Many scholars think that John knew Mark's gospel.

he had actually said it. Matthew felt the difficulty and toned it down into the innocuous, 'Why do you ask me about what is good?' (Matt. 19.16), though Luke was made of sterner stuff and retained the saying (Luke 18.19). However, Luke apparently could not stomach that desolate word from the cross: 'Eloi, Eloi, lama sabachthani? . . . My God, my God, why hast thou forsaken me?' (Mark 15.34), and Matthew turned the Aramaic into Hebrew (Matt. 27.46), so that it became a scriptural quotation rather than a cry of dereliction.

Then there are shocking things that Jesus said: 'Leave the dead to bury their own dead' (Matt. 8.22, par.), a remark in flagrant violation of Jewish sacred duty to parents and, indeed, of all pious opinion in the ancient world,[7] and thus gratuitously offensive to readers of the gospel. There are the difficult verses in which Jesus is recorded as implying a fulfillment in the lifetime of his hearers which, in the most immediate sense of his words, did not come about in any obvious fashion (Mark 9.1; 13.30, par.). When Matthew wrote of Jesus as sending out the twelve with the words, 'You will not have gone through all the towns of Israel, before the Son of man comes' (Matt. 10.23), the evangelist was as aware as we are that that saying did not find an obvious literal fulfillment. Yet he put it in his gospel.[8]

* * *

These internal considerations of the gospel texts encourage in me the view that their authors were people who were attempting to achieve a genuine historical authenticity in their accounts of Jesus. They were concerned to tell it like it was, within the conventions of their time. They would have been making use of the oral tradition transmitted in the generation that intervened between the crucifixion and the writing of Mark (a gap equivalent to that between the present day and my own first encounters with leaders of the high energy physics community, of which I believe I retain a lively and essentially dependable reminiscence). The world of their day was one which cultivated the art of oral transmission, and there are reasons to suppose that there might also have been written records now lost, such as the hypothetical, but plausible, Q, still held by many scholars to lie behind the sayings material common to Matthew and Luke.

---

[7] E. P. Sanders, *Jesus and Judaism* (London: SCM Press, 1985), 252–4.

[8] The preceding section is from *The Faith of a Physicist* by John Polkinghorne © 1994 by John Polkinghorne, published in the USA by Princeton University Press, reprinted by permission of Princeton University Press, 88–91.

Of course, one does not deny that the evangelists moulded their material, by selecting it and ordering it. It is instructive to compare the versions of the Lord's Prayer (Matt. 6.9–13; Luke 11.2–4) and the Beatitudes (Matt. 5.3–11; Luke 6.20–23) presented by two gospel writers. There is basic agreement on substance but significant variation of emphasis and elaboration. Redaction criticism has cast helpful light on this process.[9] Nor can one deny that the early Christian communities sometimes felt able to create prophetic words which spoke to their circumstances and which were then incorporated into the tradition as if they had been uttered by the earthly Jesus.

A passage like Matthew 18.15–17, which gives detailed instructions about how to deal with sinners in the 'church' (*ecclēsia* only occurs here and in Matt. 16.18 in all the gospels) quite obviously arose in this way. Yet there were equally obviously limits to this process. Acts and the Pauline epistles make it clear how critical a question it was in the early Church whether Gentiles should be circumcised or not, but no one invented a word of Jesus to settle the issue. Paul's discussion of marriage (1 Cor. 7.8–16) makes it plain that he distinguishes between what he believes originates from the Lord and what is his own apostolic judgement. Sometimes one may suppose that development took place which drew out the implications of something Jesus had said, amplified in the light of further experience. He can scarcely have been as explicit about food regulations as Mark 7.14–23 par. suggests – for otherwise how could the disputes have arisen, to which Acts and the Pauline epistles testify? – and clearly Mark 7.19b ('Thus he declared all foods clean') is an editorial gloss underlining the message. Yet it is not inconceivable that the original enigmatic saying about defilement coming from inside rather than outside was in fact Jesus' utterance.

Scholars have sometimes sought to sift the words of Jesus from later constructions by having recourse to the criterion of 'double dissimilarity.' A saying is considered authentic if it expresses a view distinct both from contemporary Jewish thought (as far as that is known) and from the concerns of the early Church. While one can concede the positive force of this criterion, to use it as the sole test of authenticity is clearly absurd. It would produce a Jesus entirely without anchorage in his society and bereft of lasting influence upon his successors. Applied to a scientist, it would assign to Schrödinger his idiosyncratic and unsuccessful attempts

---

[9] See, for example, E. P. Sanders and M. Davies, *Studying the Synoptic Gospels* (London: SCM Press, 1989), ch. 14.

at unified field theories but regard his celebrated quantum mechanical wave equation as being of questionable attribution.

It seems to me that questions of authenticity cannot be settled simply by the application of quasi-algorithmic evaluative procedures, like double dissimilarity, but that one has to take the risk of relying on the tacit skill of judgement. Biblical studies must resist the temptation to devise procedures which ape a superficial notion of scientific method. Rather, it should have the intellectual nerve to proceed in its own proper way. This will require a combination of critical assessment with empathetic interpretation. I cannot feel altogether happy with Sanders and Davies' assertion that 'The basic means of establishing evidence is cross-examination. The gospels must be treated as "hostile witnesses" in the court room.'[10] It seems to me that we must take the risk of a more subtle and open approach.

The first question, surely, is whether the gospels give the impression of having behind them a powerful personality whose character we can, at least partially, discern. I believe the answer to be yes. Just to take one example, consider the question of the parables. It is extraordinarily difficult to make up tales which have the penetrating power and haunting quality which the stories of the prodigal son (Luke 15.11–32), or the good Samaritan (Luke 10.29–37), or the parable of the sheep and goats (Matt. 25.31–46) possess. Yet some scholars are willing to suppose those small early Christian communities to have been well-endowed with creative story-tellers of this remarkable kind. It seems to me much more likely that there is one outstanding mind behind most of the parables. Once again, that is not to deny the subsidiary role of the early Church in moulding and using the material. The reference to the nations (*ethnē*) in Matthew 25.32 may well originate in the context of a Gentile mission, rather than the ministry of Jesus, which seems well-attested as being directed towards Israel.

The acuteness in controversy, by which the questioner so often has the question turned round to reveal his own presuppositions and consequent prejudice (see, for instance, healing on the sabbath (Mark 3.1–6, par.), authority (Mark 11.27–33, par.), and the tribute money (Mark 12.13–17, par.)), again seems to me to be the record of the action of a distinct and incisive mind.

Thus I find the agnosticism of some scholars about what can be known of Jesus to be most unconvincing. Talk of the 'largely unknown man of Nazareth'[11] seems to me to be grotesque. I would rather say with Pannenberg,

---

[10] Ibid., 301.

[11] J. Hick, *Death and Eternal Life* (London: Collins, 1976), 117.

'It is quite possible to distinguish the figure of Jesus himself, as well as the outlines of his message, from the particular perspective in which it is transmitted through this or that New Testament witness.'[12] The considered judgement of C. H. Dodd was that 'the first three gospels offer a body of sayings on the whole so consistent, so coherent, and withal so distinctive in manner, style and content, that no critic should doubt, whatever reservations he may have about individual sayings, that we find reflected here the thought of a Single, unique teacher.'[13] Let me add one more voice to show that I take my amateur stand in the company of scholars of great learning and unquestioned integrity. Charlie Moule summarizes a study of the tradition by saying:

> the general effect of these several more or less impressionistic portraits is to convey a total conception of a personality striking, original, baffling yet illuminating. And it may be argued that it is difficult to account for this except by postulating an actual person of such a character. The very fact that the total impression is made up of several different strands of tradition, originating (one may reasonably presume) in different circles, compensates in some degree for the absence of any rigorous test by which the authentic and original has been isolated within any one strand of tradition. If all of them, for all their diversity, combine to create a coherent and challenging impression, this is significant.[14]

Yet I must also concede that there are aspects of the gospel picture which are uncongenial to the modern reader. We need not accept Albert Schweitzer's exaggerated depiction of an apocalyptic figure trying to force a divine turning of the wheel of history by his self-immolation, but the note of urgent choice in the face of the imminent arrival of a new age is certainly there (Mark 10.29–30, par.; Matt. 8.11–12, par.; etc.), as is the idea of an eschatological trial prior to the End (Mark 10.38, par.; Mark 13, par.). Though these themes may have been elaborated by the early Christians, it is difficult to think that something like them was not also part of Jesus' view. He is thoroughly embedded in the thought patterns of his age. If he were truly human, what else could he be? Yet his concern is with God's judgement and mercy, not with datable prediction (Mark 13.32, par.).

Much more disturbing are those passages in which Jesus is portrayed as speaking in scathing, or even vituperative, terms of his opponents

---

[12] W. Pannenberg, *Jesus: God and Man* (London: SCM Press, 1968), 23.

[13] Dodd, *The Parables of the Kingdom*, 33.

[14] C. F. D. Moule, *The Origin of Christology* (Cambridge: Cambridge University Press, 1977), 156.

(Mark 12.38–40, par.; Matt. 23.13–36; etc.). Perhaps some of the language reflects the later tension between church and synagogue, but we must be wary of making hard sayings easy by attributing all the disagreeableness to the early Christians. There must have been a forceful sternness about Jesus which makes the cleansing of the Temple a credible episode. There is a puzzling complexity, as well as an attractiveness, about the figure of Jesus which drives those who speak of him to use words like 'baffling' or 'mysterious.'

One final general point must suffice. The New Testament writings are, on anyone's view, a remarkable body of religious literature. The Christian Church which has sprung from them has proved a persistent and fruitful movement, however ambiguous aspects of its history may be. These are notable phenomena, whose earliest witnesses consistently relate their experience to an origin expressed in terms of Jesus. To say the least, it would seem reasonable to explore the possibility that in this man is to be found the clue to the interpretation of what followed him, rather than locating it in the supposed creativity of a community which somehow conjured such richness from the obscure recollection of a dim and shadowy figure preceding them. In Martin Hengel's words, 'Even the most radical sceptic cannot avoid the simple historical question how this simple wandering teacher and his outwardly inglorious death exercised such a tremendous and unique influence that it still remains unsurpassed.'[15,16]

* * *

One of the points on which there is most scholarly agreement is that Jesus proclaimed the kingdom or rule of God: 'The time is fulfilled, and the kingdom of God is at hand; repent, and believe in the gospel' (Mark 1.15). His words and deeds are part of that manifestation (Matt. 12.28, par.; Matt. 13.16–17, par.; etc.). Many of the parables, including some otherwise hard to understand (e.g., the unjust steward, Luke 16.1–8), seem to have as their point the urgency of responding to this outbreaking of divine rule in the end-time of history. The offer of the kingdom is there for those able to receive it: 'Blessed are you poor, for yours is the kingdom of God' (Luke 6.20, par.). The New Testament usage of the phrase is overwhelmingly

---

[15] M. Hengel, *Atonement* (London: SCM Press, 1981), 72.

[16] The preceding section is from *The Faith of a Physicist* by John Polkinghorne © 1994 by John Polkinghorne, published in the USA by Princeton University Press, reprinted by permission of Princeton University Press, 91–5.

contained in the synoptic gospels, which strongly suggests its authenticity on the lips of Jesus.

E. P. Sanders says, 'It used to be almost a taboo question in many circles to ask what Jesus was up to.'[17] He believes this question is answerable and that one should start with some of the acts of Jesus, which Sanders thinks can be established with greater certainty than his words. 'Almost indisputable facts' are that Jesus was baptized by John, that he was a Galilean who preached and healed, called disciples and spoke of a group of twelve, confined his activity to Israel, was involved in controversy about the Temple and was crucified outside Jerusalem by the Romans. To these facts, Sanders adds that the movement around Jesus persisted after his death and that it was to some extent persecuted by the Jews.[18] At the end of his discussion Sanders gives a list of further conclusions ranging from the 'certain or virtually certain' to the 'conceivable,' which place Jesus within the Jewish contemporary setting as Sanders describes it (broadly 'restoration eschatology,' centring on the hope of a new Temple, and 'covenantal nomism,' the community of grace within the Law), while emphasizing that Jesus took an unprecedented line in accepting sinners within the kingdom of God without prior insistence on repentance. Sanders also adds a list of some 'incredible' conclusions, rejecting any idea which would make the Pharisees implacably legalistic and give to Jesus a monopoly of the concepts of love and mercy.[19] It is clear that a perspective is here imposed on the selection of gospel material to accord with Sanders' learned reconstruction of first-century Palestinian Judaism.[20,21]

* * *

We return to the gospels, therefore, seeking from them more than a skeletal, minimal portrait, and with the expectation that it should prove possible to detect, in Moule's words, 'a personality striking, original, baffling yet illuminating,' who might indeed credibly have been the trigger for the great Christian explosion into history.

---

[17] Sanders, *Jesus and Judaism*, 2.

[18] Ibid., 11.

[19] Ibid., 326–7.

[20] See also E. P. Sanders, *Paul and Palestinian Judaism* (London: SCM Press, 1977).

[21] The preceding section is from *The Faith of a Physicist* by John Polkinghorne © 1994 by John Polkinghorne, published in the USA by Princeton University Press, reprinted by permission of Princeton University Press, 95–6.

I think that a good way into the quest is through the attempt to identify traces of the *ipsissima vox* of that striking personality, looking for tones and phrases which by their vividness suggest the impress of a well-remembered, oft-recounted character.[22] I have already suggested that the haunting nature of some of the parables, and the penetrating challenges to prejudged suppositions, carry for me just that note of encounter with a highly original and individual mind. Another clue is provided by those untranslated Aramaic words which occasionally turn up in the Greek text of the New Testament. Much the most frequent of these is *amen*. First-century Jews would have used this as we do today, at the end of a prayer to signify assent. Jesus' usage, however, is entirely different. He places it at the beginning ('Amen, I say to you . . .'), where its force is to assert the unshakeable certainty of what is to follow. This occurs frequently in all four gospels – in John, in the even more emphatic reduplicated form 'Amen, amen . . .' It is a pity that so many versions disguise this fact by using translations such as 'truly,' which fail to acknowledge the singularity of the underlying word. I personally think that this idiosyncratic usage is a preserved reminiscence of a characteristic form of speech and that it testifies to one who was conscious of possessing a particular and true insight, so that he 'taught them as one who had authority, and not as the scribes' (Mark 1.22, par.).[23] It is indeed believable that he spoke as Matthew portrays him in the Sermon on the Mount (Matt. 5), saying, 'You have heard that it was said to the men of old . . . But I say to you . . .' It is the *Torah*, the Mosaic Law, which is being deepened or even corrected (Matt. 5.32, 39, 44) in these passages. Here is no rabbinic commentator, offering an interpretative gloss, but instead someone who is claiming for himself an authority capable of being set alongside the divine revelation from Sinai. The formula is not even the prophetic, 'Thus says the Lord . . . ,' but, 'This is what I say . . .' The implications of that bear some pondering. Jesus is also portrayed as exercising an authority, perceived by critics as amounting to the usurping of divine prerogative, when he pronounced the forgiveness of sins (Mark 2.5–9, par.; Luke 7.48–49).

Another significant Aramaic word is *abba*, an intimate family word for father, with more than a trace of 'daddy' to it, used as an address to God.[24] It only occurs once in the gospels, in the fraught setting of Jesus' prayer

---

[22] J. Jeremias, *New Testament Theology*, vol. 1 (London: SCM Press, 1971), ch. 1.

[23] Ibid., 35–6.

[24] Ibid., 61–8, but cf. G. Vermes, *Jesus the Jew* (London: SCM Press, 1983), 210–13.

in the Garden of Gethsemane (Mark 14.36), where it is immediately translated 'Father' (*patēr*). The retention of the Aramaic together with the Greek
suggests that here is another significant reminiscence, an impression
strengthened by the continued use of *abba* in the early tradition (Rom.
8.15; Gal. 4.6). It would seem to speak of a peculiarly close relationship
that Jesus experienced with God, something altogether deeper and more
direct than that expressed in the respectful and more distant address which
would have been contemporary Jewish practice.

The next phrase to consider is one on which there has been perhaps
more discussion and disagreement than almost any other in the New
Testament. It is 'the Son of man.' Although the equivalent Aramaic phrase
(*bar nash(a)*) never appears transliterated in the New Testament, the expression is so odd in Greek (as odd, in fact, as it is in English) that it is clear
that it is an Aramaicism which is being reproduced. Three kinds of consideration have to be taken into account in seeking an understanding of
the New Testament usage.

Firstly, 'son of man' is a perfectly normal semitic way of articulating
what we would more simply express through the single word 'man.' The
Old Testament furnishes many examples of this use of the equivalent
Hebrew phrase *ben adam* (e.g., Ps. 8.4 and Ezekiel *passim*). In the New
Testament, however, the predominant usage is to employ the definite
article in the Greek, encouraging the translation 'the Son of man.' This fact
has been emphasized particularly by Moule,[25] who takes it as indicating
that reference is being made to a particular figure, rather than humanity
in general. The natural referent is the 'one like a son of man' of Daniel
7.11–18, a heavenly figure involved in the vindication before God of the
persecuted 'saints of the Most High.' Later apocalyptic literature (the Book
of Enoch, of disputed date in relation to the Christian era) took up this
figure and developed it. In contrast to this interpretation, Geza Vermes
believes that *bar nash(a)* was commonly used in first-century Aramaic as
a kind of modest circumlocution referring to oneself,[26] rather like the
English upper-class usage, 'one likes a gin and tonic before dinner.' Once
again it is a disputed scholarly point how well this practice has been
established outside the gospels.

The second consideration relates to the way in which the phrase is actually employed. Sometimes it is the equivalent of 'man' (the sabbath saying
of Mark 2.28, par.); sometimes it seems to carry Vermes' circumlocutory

---

[25] Moule, *The Origin of Christology*, 11–22.
[26] Vermes, *Jesus the Jew*, 163–8.

sense (Matt. 16.13, cf. the parallels in Mark 8.27 and Luke 9.18, where 'Son of man' is replaced by 'I'; see also Matt. 8.20, par.). Sometimes it carries a clear apocalyptic reference and it appears to be associated closely with Jesus (the passion predictions (Mark 8.31, par.; etc.); the reply to the high priest in Mark 14.62, par.). At other times the overtones of Daniel are less clear and the connection with Jesus more problematic. 'For whoever is ashamed of me and of my words in this adulterous and sinful generation, of him will the Son of man also be ashamed, when he comes in the glory of his Father with the holy angels' (Mark 8.38, par.). Here there is an intimate connection between the attitude to Jesus and the attitude of the Son of man, but it is not at all clear that the two figures are identified.

Thirdly, all four gospels frequently attribute the phrase to Jesus, but in only one trivial case (John 12.34) is it ever on the lips of anyone else. Outside the gospels, it is used only once, by Stephen in Acts 7.56, in relation to a heavenly vision, and twice without the article in Revelation (1.13; 14.14), where images of Daniel 7 are clearly being used in connection with visionary experience.

Such are the facts. What are we to make of them? One suggestion is that 'the Son of man' was a title for Jesus employed by the very early Church, after its having been culled from Daniel 7, and that it was then read back into his earthly life when the gospels were composed. The lack of any attestation of that primitive usage makes it very unlikely, in my judgement, that the phrase is a Jewish-Christian invention. If it came from the post-Easter community, it is incomprehensible that a verse like Mark 8.38 should exhibit any ambiguity about the identification of Jesus with the Son of man. The persistence of the phrase in the gospel tradition (from Mark to John, if that is the chronological order of their writing), and its virtual absence elsewhere, seem much more probably explained by a correctly remembered attribution to Jesus himself. I am persuaded by Moule's arguments to see a significant reference to Daniel 7, so that the Son of man was for Jesus 'a symbol of a vocation to be utterly loyal, even to death, in the confidence of ultimate vindication in the heavenly court.'[27]

What then of that unclear conjunction hinted at in Mark 8.38, between the heavenly figure and Jesus himself? Here one must venture down the way, persistently placarded with warning signs by cautious scholars, of attempting some estimate of Jesus' self-understanding. The reason for the care needed is that this is an area where one is most vulnerable to the

---

[27] Moule, *The Origin of Christology*, 14.

presence in the sources of the retrospective effects of later Christian hind-sight. Nevertheless, the attempt must be made if we are to succeed in our aim of finding a credible picture of the one who was the initiator of the Christian movement. 'Nothing comes of nothing,' and so much has come of Jesus that there must have been something present in him in the first place which was commensurate with the effect it produced. I think New Testament theologians in their attempts to reconstruct Christian origins need something of the corrigible boldness displayed by cosmologists in their approach to a similar task.

There are two extremes to be avoided. One is to attribute to Jesus such extraordinary powers that he effectively ceases to be credibly a recogni-zable human being. The other is so to recoil from this error that one treats him as if he were an uninteresting mediocrity. A good test case is pro-vided by the so-called 'passion predictions,' whose status is important in relation to an assessment of Jesus' self-understanding and purpose. Three times in each synoptic gospel, Jesus is portrayed as telling his disciples that he will be rejected and killed and after three days rise again (Mark 8.31; 9.31; 10.33–34; and par.). Many scholars regard these passages as prophecies-after-the-event, inserted by the post-Easter Church. I do not doubt that their form has been influenced by retrospection. I cannot think that Jesus saw his future laid out before him in fine detail, for I believe he lived a truly human life to which such precise foreknowledge would be foreign. But equally I cannot believe that he did not see in general terms that rejection and execution awaited him in Jerusalem, or that he did not trust that nevertheless he would be vindicated. To believe less than that is to make him out to be lacking in insight and faith.

The woodenness of some New Testament scholarship is apparent when it is suggested that there is some tension between what I have just said and the story of the agony in Gethsemane (Mark 14.32–42, par.), as if foresight dissolves the intensity and demand of the actual moment. Equally lacking in imagination is the suggestion that the story of the Garden must be made up because the sleeping disciples could not have known what was going on, as if one could not conceive of a restless panicky dozing – half flight from awful reality, half inescapable consciousness of it. This deeply moving but conventionally unheroic story (compare it with the death of Socrates or the deaths of many subsequent Christian martyrs) must, I believe, be a true reminiscence and of some profound significance. It is also attested elsewhere, outside the gospel tradition (Heb. 5.7).

The basis of Jesus' understanding of his mission lay in his firm confidence in God his Father, not in a detailed foreknowledge of what would happen.

I believe that this view is consistent with Mark 8.38's assertion of the central role of Jesus himself, while leaving a certain openness as to the form of the 'ultimate vindication in the heavenly court.' The point of view I am proposing envisages that Jesus was clear as to his having received a unique call from God (the story of the baptism by John is important here), but it allows him to find the nature of its fulfilment through its unfolding realization (a process pictured in the gospels as beginning with the temptations in the wilderness, immediately subsequent to the baptism). Such a notion of an evolving vocation is not foreign to the New Testament. One thinks of Luke (2.52) and of the writer to the Hebrews, who combines a high Christology (Heb. 1.1–4) with a frank acknowledgement of development (Heb. 5.8–10).

If this picture has elements of truth in it, one might expect to find that the evocative openness of interpretation characteristic of the parables, and of that polyvalent phrase 'the Son of man,' might be found elsewhere in the utterances of Jesus, as a means of indicating an envelope of understanding within which his nature and destiny could be contained and explored. This encourages me to think that the gospel language about Jesus which is reminiscent of Old Testament concepts of God's Wisdom (Prov. 8.22–31, etc.) – language which we find in passages like Matthew 11.28–30 ('Come to me all who labour and are heavy laden . . .') and, most strikingly, in Luke 11.49 ('the Wisdom of God said, "I will send them prophets and apostles . . ."'), where in the parallel passage in Matthew (23.34) 'Wisdom' is replaced by 'I' – while perhaps moulded by later Christian thought, has an anchorage in the words and understanding of Jesus himself. Jeremias has argued from a consideration of the probable underlying Aramaic[28] that the 'Johannine thunderbolt from a clear synoptic sky' of Matthew 11.27, par.: 'no one knows the Son except the Father, and no one knows the Father except the Son and anyone to whom the Son chooses to reveal him,' is really a pithy parable (only a father really knows a son and a son his father) rather than an absolute Christological assertion. It seems to me that the choice of interpretation may not be that stark and clear cut, for this might be a case where the creative ambiguity of a parabolic saying was being used in an exercise of exploration, and that the words are neither simply general nor specifically assertive, but rather an allusive hint of what might be. I find the same heuristic possibility in the 'ransom' saying of Mark 10.45. Jesus is portrayed as one who demands a committed response, centred on himself (Matt. 8.21–22, par.;

---

[28] Jeremias, *New Testament Theology*, 56–61.

10.37–39, par.). In this way one can see the conceivability (I believe, the likelihood) that there are present in the synoptic gospels seminal sayings of the historical Jesus, which in the light of post-Easter reflection could lead to the Christ of the Johannine discourses. What is being suggested is that Jesus' self-understanding was consistent with later incarnational reflection, without being the same as it. It is also being suggested that Jesus was not curiously unreflective about himself, but that he sought images of adequate self-understanding and that he expressed them in ways which were remembered.

If elements both human and divine meet in Christ, the preservation of the true humanity surely must mean that the historical person did not go around thinking of himself as God. Something altogether more nuanced than that would have been necessary. Even in John we find this is recognized, for there are both claims of strong identification ('I and the Father are one,' John 10.30) and of clear distinction ('for the Father is greater than I,' John 14.28). In all the gospels, Jesus is portrayed as one who needed to pray. James Dunn summarizes the conclusion of his own careful survey of these questions by saying:

> We cannot claim that Jesus believed himself to be the incarnate Son of God; but we can claim that the teaching to that effect as it came to expression in the later first-century Christian thought was, in the light of the whole Christ-event, an appropriate reflection on and elaboration of Jesus' own sense of sonship and eschatological mission.[29]

Concern with these historical questions is an inescapable task for a bottom-up thinker. He cannot accept Kierkegaard's celebrated assertion that 'If the contemporary generation had left nothing behind them but these words: "We have believed that in such and such a year God has appeared among us in the humble figure of a servant, that he lived and taught in our community and finally died", it would be more than enough.'[30] Without further evidence, how could one know that this was even conceivably true? Rather, I must ask with Leonard Hodgson (in his Gifford Lectures for 1955–7), 'What must the truth have been if it appeared like this to men who thought like that?'[31]

While I acknowledge the importance of the Church's testimony to its Lord, I cannot accept a primacy of the preached Christ over the historical Jesus, of the kerygma over history. Rather, I feel impelled to strive for a

---

[29] J. D. G. Dunn, *Unity and Diversity in the New Testament* (London: SCM Press, 1980), 254.

[30] Quoted in J. Macquarrie, *Jesus in Modern Thought* (London: SCM Press, 1990), 237.

[31] Quoted in Moule, *The Origin of Christology*, 6.

mutually consistent understanding of them both. No doubt, in our encounter with Jesus there is recognition as well as surprise, so that, in Macquarrie's words, 'We recognize the historical Christ as revelation because we already have in our constitution as human beings an ideal archetype which, we believe, we see fulfilled in him.'[32] Yet that seems only half the story, for Jesus would not seem at times so strange and mysterious if only recognition were involved. Because I believe that we see in Jesus much more than an encouraging or illuminating exemplar, so that he opens up the availability of a relationship between God and humanity previously undreamed of, I could not for a minute go on to contemplate with Macquarrie the possibility that 'there may come a time when only the pure archetype will remain as the focus of a wholly rational religion, and the last links with the historical Jesus and with historical Christianity will have been snapped.'[33] In my view, Christianity will never become all 'top,' with no historical 'bottom.'

So far we have been concentrating on the words of Jesus, but his deeds are also of importance. If there was an early sayings source Q, it may well be significant that it did not survive on its own, but became incorporated into the narrative action of Matthew and Luke. The sayings source which we do have, the second-century apocryphal Gospel of Thomas, comes from a gnostic environment, where there would have been a 'top' concern with enlightened knowledge and a recoil from a 'bottom' engagement with the specificities of history.

One of Sanders' 'almost indisputable facts' is that Jesus was a healer. That would not have set him apart in the ancient world, where claims of such happenings were quite widespread, though mainly associated with cultic sites rather than individuals. The dispute about healing on the sabbath (Mark 3.1–6, par.) requires that such healings actually took place, and in general the gospel stories of healing are so numerous and interwoven into the accounts that they cannot be excised in any natural or credible way. There are a variety of ways in which to think about Jesus' mighty works. We may suppose that the psychosomatic consequence of encounter with a charismatic personality, say bringing release from hysterical paralysis, played a part in these happenings. It is possible to think of some of the nature miracles (e.g., the stilling of the storm, Mark 4.35–41, par.) as being examples of profoundly significant coincidences, such as happen from time to time. Other miracle stories resist this naturalizing tendency. It seems lame to explain away the feedings of the multitude

---

[32] Macquarrie, *Jesus in Modern Thought*, 183.
[33] Ibid., 185.

(Mark 6.32–44, par.; Mark 8.1–10, par.) by saying that an example of unselfishness induced the many to bring forth the rations they had prudently retained for their private use. Changing water into wine (John 2.1–11) is an irreducibly unnatural happening. A difficulty is that these latter stories also carry an obviously high symbolic value as signifying the difference that the presence of Jesus makes, and so they can be conceived of as tales in the tradition composed for this purpose.

Different people will judge the matter in different ways. It is impressive how matter-of-fact the gospel accounts of miracles are. Although we are often told things such as that the crowds were 'amazed and glorified God, saying, "We never saw anything like this!"' (Mark 2.12), there is little attempt to pander to a taste for the marvellous, in striking contrast to the concoctions of pious fancy found in the second-century apocryphal gospels. Only very occasionally (perhaps the coin in the fish's mouth, Matt. 17.24–7) do we find a tale of what seems mere wonder-working. Jesus' healings were not coercive interventions; they required co-operation (Mark 6.5–6, par.). Anthony Harvey has given a careful discussion in which he compares the miracle stories associated with Jesus with those current about other figures in the ancient world. He comes to 'the remarkable conclusion that the miraculous activity of Jesus conforms to no known pattern.'[34]

A judgement on miracles cannot be separated from a judgement on Jesus; there is an inescapable circularity. The more we have reason to think him exceptional, the more coherent is the possibility that he exercised exceptional powers. C. H. Dodd, who has a rather reserved attitude to the miracle stories, says that while such things do not happen in ordinary circumstances, 'the whole point of the gospels is that circumstances were far from ordinary. They were incidental to a quite peculiar situation, unprecedented and unrepeatable.'[35] It is conceivable that unexpected events occur in unprecedented circumstances. The miracle stories cannot simply be dismissed because of an a priori certainty that such things 'cannot happen,' but equally the circularity we have noted in their evaluation means that they can no longer be used simply as 'proofs' of, say, Jesus' divinity, in a way that pre-critical generations attempted (cf. Mark 8.11–12, par.).[36]

---

[34] A. E. Harvey, *Jesus and the Constraints of History* (London: Duckworth, 1982), 113.

[35] Dodd, *The Parables of the Kingdom*, 44.

[36] The preceding section is from *The Faith of a Physicist* by John Polkinghorne © 1994 by John Polkinghorne, published in the USA by Princeton University Press, reprinted by permission of Princeton University Press, 97–104.

18

The resurrection

It is absolutely clear that something happened between Good Friday and Pentecost. The demoralization of the disciples, caused by the arrest and execution of their Master, is undeniable. Equally undeniable is the fact that within a short space of time, those same disciples were defying the authorities who had previously seemed so threatening, and that they were proclaiming the one who had died disgraced and forsaken, as being both Lord and Christ (God's chosen and anointed one). So great a transformation calls for a commensurate cause. From the nineteenth century onwards, in the thought of people such as Renan and Bultmann, it has been suggested that what happened was a faith event in the minds of the disciples, a conviction achieved after a period of reflection, that the cause of Jesus continued beyond his death. Somewhat similar is the opinion of Edward Schillebeeckx, for whom the primary terms of early Christian belief about the fate of Jesus centre on his being exalted to life with God, a conviction held to arise from pondering the idea of the vindication of the righteous set out in passages such as Wisdom 2.17—3.4, and finding only secondary expression in the idea of resurrection. 'It was only when people began to see that the deliverance of Jesus was also a conquest of death itself . . . that the idea of resurrection forced itself upon all the early Christian communities everywhere as the best way of articulating the fact of Jesus' being "alive to God".'[1]

Frankly that seems to me to be a wholly unconvincing interpretation of New Testament attitudes, as subsequent analysis will seek to show. The writing with the most sustained emphasis on the theme of exaltation is the Epistle to the Hebrews, in which the language of resurrection is not used explicitly, but it is surely implicit throughout that the one who 'sat down at the right hand of God' (Heb. 10.12) was the one whom God had 'brought again from the dead' (Heb. 13.20). The notion of the exaltation of the crucified was too paradoxical a concept to have arisen in first-century Palestine simply through a process of reflection. I agree with Macquarrie

---

[1] E. Schillebeeckx, *Jesus* (London: Collins, 1974), 538.

when he says he does not think that 'reading or remembering a passage of scripture which speaks in a general way of a hope for the righteous beyond death would have been nearly percussive enough to produce in Peter and the others the radical turn-around or conversion they had at that time.'[2]

Even less likely is the theory, first suggested in the nineteenth century, and still occasionally put forward, that Jesus swooned on the cross and revived in the cool of the tomb. In addition to the many historical implausibilities involved (the Romans knew how to execute), this idea fails to convince because, as David Strauss rightly emphasized, 'It is impossible that a being who had stolen half dead out of the sepulchre ... could have given the disciples the impression he was a Conqueror over death.'[3] However, Strauss believed it was hallucinatory experience which could have been the trigger of the change in the disciples' attitude, an idea which goes back to Celsus in the second century. The remorse of Peter caused in him an abreaction after the trauma of denial, which then communicated itself to the fraught band of disciples in a psychological chain reaction; that is the way a modern person might express the thought. Such an explanation fails to account for the varieties of time and place associated in the tradition with the claimed appearances, including Paul's experience on the Damascus road which must have been three years or so after the crucifixion. It also makes a good number of unsubstantiated assumptions about the temperamental make-up of the disciples. Also, I must confess to an instinctive feeling that hallucinations, however vivid, could not have been the enduring basis of the vitality of the early Christian movement.

Sanders, in his role of cautious historian, prescinds from the issue of what went on:

> What is unquestionably unique about Jesus is the result of his life and work. They culminated in the resurrection and the foundation of a movement which endured. I have no special explanation or rationalization of the resurrection experiences of the disciples. Their vividness and importance are best seen in the letters of Paul ... We have every reason to think that Jesus had led them to expect a dramatic event which would establish the kingdom. The death and resurrection caused them to adjust their expectation, but did not create a new one out of nothing. That is as far as I can go in looking for an explanation of the one thing which sets Christianity apart from other 'renewal movements.' The disciples were prepared for *something*. What

---

[2] J. Macquarrie, *Jesus in Modern Thought* (London: SCM Press, 1990), 312.
[3] Quoted in G. O'Collins, *Jesus Risen* (London: Darton, Longman & Todd, 1987), 101.

they received inspired them and empowered them. It is the *what* that is unique.[4]

I cannot rest content with that. It seems necessary and reasonable to go on to ask, What was the explanation offered by the disciples themselves?

The New Testament answer is that they believed that Jesus had been raised from the dead and that 'To them he presented himself alive after his passion by many proofs' (Acts 1.3). It is important to remember that the earliest account of the resurrection appearances does not occur in the gospels, but in Paul's first letter to the Corinthians, written in the middle fifties AD. He had founded the Corinthian church, and he reminds them that when he did so:

> I delivered to you as of first importance what I also received, that Christ died for our sins in accordance with the scriptures, that he was buried, that he was raised on the third day in accordance with the scriptures, and that he appeared to Cephas, then to the twelve. Then he appeared to more than five hundred brethren at one time, most of whom are still alive, though some have fallen asleep. Then he appeared to James, then to all the apostles. Last of all, as to one untimely born, he appeared also to me. (1 Cor. 15.3–8)

When Paul says he delivered what he also received, it is reasonable to suppose that he is referring to teaching given immediately following his conversion, just a very few years after the crucifixion itself. Thus this testimony takes us back very close indeed to the events cited. The antiquity of the material is confirmed by the use of the Aramaic 'Cephas' for Peter, and by the reference to 'the twelve,' a phrase which soon fell out of Christian usage. The style of the reference to the five hundred brethren makes it plain that an appeal to accessible evidence is being made, an appeal which Bultmann, with his maximal distrust of the historical, regarded as dangerous, but which I, of course, whole-heartedly welcome. It is entirely possible that this attestatory role is the reason why there is no reference in Paul's list to the witness of the women (prominent in the gospel accounts of appearances), since in the ancient male-dominated world their testimony would not have been validly acceptable.

The account in 1 Corinthians 15 is extremely spare, a simple list of witnesses. It concludes with Paul's own encounter with the risen Christ, referred to again by the apostle himself in Galatians (1.11–17), and three

---

[4] E. P. Sanders, *Jesus and Judaism* (London: SCM Press, 1985), 320.

times described in Acts[5] (9.1–9; 22.6–11; 26.12–18; with minor variations between the accounts, which provide another example of the kind of degree of detailed consistency which even a single author thought it necessary to achieve in the first century). Pannenberg reminds us that Paul is the only writer whose words are certainly those of a resurrection witness.[6] One might suppose Paul's experience to be best categorized as a vision. Pannenberg uses that language, but he emphasizes that if that is so, vision is used in a distinctive sense, for elsewhere the New Testament proves perfectly capable of speaking of visionary experience in terms carrying much less weight of significance than is attached to the case of the resurrection appearances (e.g., Acts 23.11). 'If the term "vision" is to be used in connection with the Easter appearances, one must at the same time take into consideration that primitive Christianity itself apparently knew how to distinguish between ecstatic visionary experience and the fundamental encounters with the resurrected Lord.'[7]

Paul clearly places much less emphasis on a remarkable 'vision and revelation of the Lord' which he had received (2 Cor. 12.1–7; plainly a coy self-reference) than on his Damascus road encounter, which is the ground of his apostleship. 'Am I not an apostle? Have I not seen Jesus our Lord?' (1 Cor. 9.1). It is critical for his authority that he should find a place in that list of witnesses, alongside Peter and James and the rest. So his experience must be comparable to theirs. To assess what that might be and to get beyond that spare enumeration, we have to turn to the appearance stories in the gospels.

Immediately one enters a strange, almost dream-like world, in which Jesus appears in rooms with locked doors and suddenly disappears again. There is considerable difference of account between the different gospels. This latter point is in marked contrast to the preceding stories of the passion. These certainly display variations of detail (particularly in relation to the trials which Jesus underwent), but perhaps to no greater degree than one might expect in traditions stemming from recollections of a confused and frightening twenty-four hours. They are plainly recounting the same broad sequence of events. The gospel treatments of the resurrection appearances are much more diverse.

---

[5] I take a higher view of the basic historical reliability of Acts than do some New Testament scholars.

[6] W. Pannenberg, *Jesus: God and Man* (London: SCM Press, 1968), 77. This implies certain reservations about the authorship of other New Testament writings (such as John and 1 Peter).

[7] Ibid., 94.

Mark, at least as far as the authentic text available to us is concerned, does not give a description of any appearance of the risen Jesus, though one is foreshadowed in 16.7: 'he is going before you to Galilee; there you will see him, as he told you.' Scholars have argued whether there is a lost conclusion to the gospel which would have supplied the present lack. Our text ends with the words about the women at the tomb, 'they said nothing to any one, for they were afraid' (Mark 16.8). Part of the discussion has been whether it was possible for a Greek text to conclude with *gar* (for). It now seems that this is conceivable. Whether such an ending is convincingly fitting is another matter. It suits a certain modern taste to end in mystery and fear, but I greatly doubt whether a first-century writer would have seen it that way. Certainly, by the second century it was felt appropriate to construct the additions to Mark which figure in some manuscripts and many of our translations. These incorporate ancient tradition, but they cannot be considered to constitute an independent witness in relation to the other gospels.

Matthew records a meeting of the risen Jesus with the women (Matt. 28.9–10), at which he tells them he will meet his 'brethren' in Galilee, and subsequently such a meeting is described (Matt. 28.16–20). The account of the latter includes an articulated trinitarian formula which must surely be a quite late development in the tradition.

In Luke 24 everything happens in Jerusalem on the first Easter day itself. Jesus does not meet with the women, but he journeys to Emmaus with two of the disciples and later he appears to the assembled eleven, finally parting from them after he has led them out to Bethany. There is also a brief reference to an appearance to Peter ('Simon,' Luke 24.34; a verse that has about it something of the air of an oft-repeated credal statement). The same author's Acts speaks in general terms of appearances stretching over a period of forty days.

The most extensive sequence of resurrection appearances is described in John. Jesus is seen by Mary Magdalene (John 20.11–18). Then he appears to the eleven, less Thomas, on Easter evening, and a week later again to them all, including Thomas this time, who utters the most unequivocal assertion of Jesus' divinity found in the New Testament: 'My Lord and my God!' (John 20.19–29). In what appears to be an appendix added to the gospel (compare 20.30–31 with 21.24–25), we are given the detailed story of an appearance by the lakeside in Galilee (John 21.1–23), which has some elements in common with an incident which Luke locates in Jesus' lifetime (Luke 5.3–7).

It is a somewhat confusing mass of material. Pannenberg believes that 'The Easter appearances are not to be explained from the Easter faith of the disciples; rather, conversely, the Easter faith of the disciples is to be explained from the appearances.'[8] However, his estimate of the gospel material is that

> The appearances reported in the Gospels, which are not mentioned by Paul, have such a strongly legendary character that one can scarcely find a historical kernel of their own in them. Even the Gospels' reports that correspond to Paul's statements are heavily coloured by legendary elements, particularly by the tendency toward underlining the corporeality of the appearances.[9]

I agree with the judgement about appearances leading to faith, but not with the assignment of almost all detail to the category of legend. Amid the variety of the appearance stories there is one element which is both unexpected and persistent. It is that there was difficulty in recognizing the risen Jesus. Dodd is right to say that the stories 'are all centred in a moment of recognition,'[10] and to use that as an argument against their being assimilated to a category of vague mystical experience, but that moment of recognition is not easily attained. Mary Magdalene mistakes Jesus for the gardener; on the Lake of Galilee, only the beloved disciple has the insight to recognize that the figure on the shore is the Lord; the couple walking to Emmaus only realize who their companion has been at the moment of the breaking of bread and his disappearance; with great frankness Matthew tells us that when Jesus appeared on the mountain in Galilee 'they worshipped him; but some doubted [*edistasan*]' (Matt. 28.17). This would be a strange motif to recur in stories which were merely made up. It seems likely to me that, on the contrary, it is the kernel of a genuine historical reminiscence.

The corporeality claimed for the risen Jesus is emphasized in Luke, here Jesus encourages the disciples to handle him, 'for a spirit has not flesh and bones as you see that I have' (Luke 24.39), and where he eats some broiled fish; and to a lesser extent in John where, though Mary Magdalene is told not to cling to him, the disciples are invited to inspect the wounds, and where, by implication, Jesus might be thought to have joined in the lakeside meal.[11] What one makes of this – whether one automatically reaches for the category 'legendary' – depends upon one's understanding

---

[8] Ibid., 96.

[9] Ibid., 89.

[10] C. H. Dodd, *The Parables of the Kingdom* (Welwyn: James Nisbet, 1961), 40.

[11] One might ask the question, Did the risen Christ breathe? That too would involve exchange between his glorious body and the environment.

of human embodiment and its conceivable destiny. Commenting on the Lukan passage, G. B. Caird says that 'to a Jew a disembodied spirit could only seem a ghost, not a living being, but a thin, unsubstantial carbon-copy which had somehow escaped from the filing system of death.'[12] J. I. H. McDonald says that 'Luke was ruling out any truck with notions such as subjective vision, psychic peculiarity or insubstantial shade to account for the risen figure of Christ. Jesus is real and is found among the living.'[13] I am very wary of those who want to take too exclusively spiritual a view of anything relating to humanity. For that reason I do not warm to those who use abstraction from the Acts accounts of the Damascus road experience to argue that luminous glory, rather than a focus on personal identity, lies behind the appearance stories. This led J. M. Robinson to claim that the vision of the exalted Christ experienced by the seer of Patmos (Rev. 1.9–20) is 'the only resurrection appearance in the New Testament that is described in any detail.'[14] Commentators have sometimes suggested, in a similar vein, that the story of the transfiguration (Mark 9.2–8, par.) is a displaced resurrection appearance. These seem to me to be implausible manipulations of New Testament material, whose authors should be given greater credit for knowing what they were doing. As J. A. Baker points out, the actual appearance stories in the gospels conform to neither of the contemporary models for post-mortem phenomena, which are a dazzling heavenly figure or a resuscitated corpse.[15]

Acknowledgement of a degree of corporeality in the appearance accounts is far from equating the resurrection with a mere resuscitation. Whatever we may make of the stories of those whom Jesus restored to life (Mark 5.21–43, par.; Luke 7.11–17; John 11.1–44), there is no question that they were destined eventually to die again. They were resuscitated, not resurrected. Jesus, however, is raised to endless life; his resurrection body is transmuted and glorified, possessing the unprecedented properties that allow him to appear and disappear in locked rooms, yet bearing still the scars of the passion. What such continuity and discontinuity might mean is tentatively explored by Paul in 1 Corinthians 15 (vv. 35–50) in terms of our own eventual destiny beyond death. He warns his readers against a resuscitatory reductionism ('flesh and blood cannot inherit the kingdom of God, nor does the perishable inherit the imperishable,' v. 50). Yet the

---

[12] G. B. Caird, *The Gospel of Luke* (New York: Penguin, 1963), 261.

[13] J. I. H. McDonald, *The Resurrection* (London: SPCK, 1989), 107.

[14] Quoted in O'Collins, *Jesus Risen*, 212.

[15] J. Baker, *The Foolishness of God* (London: Darton, Longman & Todd, 1970), 253–5.

tone of the passage is also against a spiritual reductionism, for it is the resurrection *body* which is being discussed. It is necessary to consider a second line of evidence which has a potential bearing upon such questions. I refer, of course, to the stories of the empty tomb.

All four gospels contain accounts of how, once the sabbath was over, women came to the tomb to attend to the body of Jesus, only to find the stone rolled away and the tomb empty (Mark 16.1–8, par.). The stories differ in details of timing in relation to dawn, the names of the women involved and the number of angelic messengers they encountered. However much these discrepancies may have disturbed Ernest Pontifex in the atmosphere of narrow literalism portrayed in *The Way of All Flesh*, they are unlikely to disconcert us. We can accept such variation without believing that this by itself casts doubt on the core tradition.

For some, these stories are the strongest evidence for the resurrection. Why did the Jerusalem authorities not nip the nascent Christian movement in the bud by exhibiting the mouldering body of its leader? It is incredible to suggest that the disciples stole the body in an act of contrived deceit, and unbelievably lame to suggest that the women went to the wrong tomb, so that it all arose from a mistake. The only credible reason for the emptiness of the sepulchre was that Jesus had actually risen. So the argument goes.

A somewhat more careful assessment is required. The first explicit account of the empty tomb is in Mark, written some 35 years or so after the event. It is suggested by some scholars that we have here a second-generation story, made up as the expression of an already existing conviction (perhaps based on the appearances) that Jesus had survived death. Even the fact of a separate tomb at all is held to be questionable, for it was the common Roman practice to inter executed felons in the anonymity of a common grave. A number of points may be made in response.

While it is notorious that Paul does not refer explicitly to the empty tomb in his extant letters, not only is the argument from silence particularly dangerous when applied to such occasional writings, but also the occurrence of the phrase 'was buried' in that extraordinarily spare summary in 1 Corinthians 15 seems clearly to indicate that a special significance attached to the burial of Jesus. It seems very hard to believe that a Jew like Paul, whose background of thought would have been one emphasizing the psychosomatic unity of the human being, could have believed that Jesus was alive but that his tomb still contained his mouldering body. James Dunn concludes a survey of first-century Pharisaic thought and practice by saying, 'the ideas of resurrection and of empty tomb would

naturally go together for many people. But this also means that any assertion that Jesus had been raised would be unlikely to cut much ice unless
his tomb was empty.[16]

There is archaeological evidence from Palestine later in the first
century which shows that a crucified man was, in that case, buried separately and not assigned to a common grave. Thus the story of Jesus'
separate burial is not impossible. If it were a made-up story, it is hard to
see why Joseph of Arimathea and Nicodemus are the names associated
with it, since these figures do not play any prominent part in the subsequent story of the Christian movement. The most natural explanation of
their assignment to an honoured role is that they fulfilled it.

Equally, if the discovery of the empty tomb were a concocted fiction,
why, in the male-dominated world of that time, were women chosen to
play the key parts? Far and away the most natural answer is that they
actually did so. Of course, there are oddities about the story. How did the
women imagine they were going to cope with the heavy stone blocking
the entrance? (This problem is acknowledged in the account: Mark 16.3.)
After three days, in that hot climate, would it not have been too late to
attend to the corpse? However, contemporary understanding held that
corruption set in on the fourth day (cf. John 11.17, 39). John alone suggests that some preliminary precautions had been taken on the Friday
evening (John 19.39–40). Such problems are, perhaps, more characteristic of the roughness of reminiscence than the smoothness of composition
and, in any case, one should not expect coolly logical behaviour from
women still distraught at the execution of their revered Master.

Whatever difficulties twentieth-century scholars may feel about the
empty tomb stories, they do not seem to have been shared by critics of
Christianity in the ancient world. As a bitter polemical argument sprang
up between Judaism and the Church, it was always accepted that there
was a tomb and that it was empty. The critical counter-suggestion was
that the disciples had stolen the body in an act of deception, an explanation which I regard as incredible. Just how far back this argument can
be traced is indicated by the story of the watch set on the tomb (Matt.
27.62–66; 28.11–15). I consider this to be a patently fabricated tale from
a Christian source, concocted precisely to rebut the canard that the
disciples had been grave-robbing. There is clear evidence, then, that in the
first century those hostile to Christianity nevertheless accepted that
the tomb had been found empty. A confirmatory consideration is the

---

[16] J. D. G. Dunn, *Unity and Diversity in the New Testament* (London: SCM Press, 1985), 67.

complete lack of any evidence of a cult associated with the burial place of Jesus. Ancient Jewish piety was much given to respectful veneration of the tombs of prophets and patriarchs (cf. Matt. 23.29). The total absence of this in the case of Jesus strongly suggests that from the first it was realized that for him the tomb was an irrelevancy. Christian interest in the possible burial place only dates from later centuries, when an increasing engagement of Christian thought with history led to giving attention to sites associated with Jesus' life.

Thus there are many reasons for taking seriously the tradition of the empty tomb, in addition to the tradition of the appearances of the risen Christ. Dodd summarizes his assessment of the gospel writers' narratives by saying, 'It looks as though they had a solid piece of tradition, which they were bound to report because it came down to them from the first witnesses, though it did not add much to the message they wished to convey, and they hardly knew what use to make of it.'[17] He is emphasizing the fact, contrary to some modern apologetic strategies, that the gospels do not present the empty tomb as a knock-down argument for the truth of the resurrection. Rather, it requires explanation. Hence the need for the message of the angel, 'He has risen, he is not here' (Mark 16.6, par.); the flow of understanding is from resurrection to absence of the body, rather than the reverse. In the story of Peter and the beloved disciple at the tomb (John 20.3–10), it is only the latter who has the insight to recognize unaided what has happened. For the others, the discovery of the emptiness of the tomb is, at first, disorientating; 'a nasty shock.'[18] There is no easy triumphalism in these stories, which itself makes one the more inclined to accept them as stemming from authentic recollection. From the point of view of the New Testament, it is the resurrection which explains the empty tomb rather than the empty tomb proving the resurrection.

There are twentieth-century Jewish writers who accept the emptiness of the tomb without thereby being driven to embrace Christianity. Geza Vermes concludes: 'In the end, when every argument has been considered and weighed, the only conclusion acceptable to the historian must be ... that the women who set out to pay their last respects to Jesus found to their consternation, not a body, but an empty tomb.'[19] The orthodox Jew, Pinchas Lapide, goes further. He believes that Jesus was raised from the dead, but he does not accept him as the Messiah, let alone the incarnation of God.

---

[17] Dodd, *The Parables of the Kingdom*, 172.
[18] McDonald, *The Resurrection*, 140.
[19] G. Vermes, *Jesus the Jew* (London: SCM Press, 1983), 41.

'Thus, according to my opinion, the resurrection belongs to the category of the truly real and effective occurrences, for without a fact of history there is no act of true faith.'[20,21]

* * *

I hope I have made it clear that there is motivation for the belief that Jesus was raised from the dead (the most ancient expression is always in the passive; it is a great act of God, not a final miracle of Jesus, which is being asserted). We now have to ask the question whether the motivation provided is in fact strong enough to support the extraordinary claim being made. Such an assessment will depend upon the second movement of thought about which I spoke when I introduced the question of the resurrection. Can it make sense within a general understanding of God and his ways with humanity that, alone of all who have ever lived, this man was restored to unending life in an act which, although it transcends history, nevertheless is embedded in history?

Inevitably we return to the hermeneutical circle in which the significance of Jesus and the truth of his resurrection inextricably interact with each other. The modern theological writer who has expressed this most clearly and emphatically is Wolfhart Pannenberg. He says bluntly, 'Jesus' unity with God was not yet established by the claim implied in his pre-Easter appearance, but only by his resurrection from the dead,'[22] and later he asserts that 'Apart from Jesus' resurrection, it would not be true that from the very beginning of his earthly way God was with this man.'[23] It is the resurrection which makes it plain who Jesus is. Paul says that 'if Christ has not been raised, then our preaching is in vain and your faith is in vain' (1 Cor. 15.14). A top-down thinker like Macquarrie demurs: 'I doubt very much whether in the case of such a complex system of beliefs as Christianity, such a simplistic mode of falsification is possible.'[24] We bottom-up thinkers view things differently, for we are open to the possibility of critical events on which an understanding pivots. For sure, once a complex physical theory, such as special relativity, has achieved 'well-winnowed' status, it

---

[20] P. Lapide, *The Resurrection of Jesus* (London: SPCK, 1984), 92.

[21] The preceding section is from *The Faith of a Physicist* by John Polkinghorne © 1994 by John Polkinghorne, published in the USA by Princeton University Press, reprinted by permission of Princeton University Press, 109–18.

[22] Pannenberg, *Jesus: God and Man*, 53.

[23] Ibid., 341.

[24] Macquarrie, *Jesus in Modern Thought*, 406.

will not be falsified by the first claim of an adverse experiment, but special relativity would never have come into being at all if Michelson and Morley had measured a non-zero velocity of the Earth through the aether in their famous experiment. I am not claiming that the whole of a developed traditional Christology can be read out from the resurrection, but I do believe that if we cannot make the claim 'Jesus lives,' the ambiguity of his death remains an unresolved enigma and the significance of his life and message seem at most a brave gesture in a hostile world. Christianity would not have come into being without the resurrection of Jesus.

Pannenberg says that 'if the cross is the last thing we know about Jesus then – at least for Jewish judgement – he was a failure.'[25] Moltmann concurs: 'As a merely historical person he would long have been forgotten, because his message had already been contradicted by his death on the cross.'[26] It seems to me entirely possible that if Jesus had not been raised from the dead we would never have heard of him. Lapide, from his Jewish perspective, speaks of 'the *must* of the resurrection': 'Jesus *must* rise in order that the God of Israel could continue to live as their heavenly Father in their hearts; in order that their lives would not become God-less and without meaning.'[27] He serves to remind us that the resurrection is not only the vindication of Jesus. It is also the vindication of God: that he did not abandon the one man who wholly trusted himself to him. Moreover, we begin to see here some glimmer of a divine response to the problem of evil. If Good Friday testifies to the reality of the power of evil, Easter Day shows that the last word lies with God. David Jenkins writes:

> We do not see how the purposes of love can be reconciled with the purposelessness of evil, but we do see that the human being who embodies the pattern of a loving God is both submerged in the destructiveness of evil and emerges from it as a distinctive, loving and personal activity. The Logos of the cosmos is not a mythological theory but a crucified man. The hope of personal sense and fulfilment lies neither in ignoring evil nor in explaining evil, but in the fact that Jesus Christ endured evil and emerged from evil.[28]

Finally, the resurrection of Jesus is the vindication of the hopes of humanity. We shall all die with our lives to a greater or lesser extent incomplete, unfulfilled, unhealed. Yet there is a profound and widespread human

---

[25] Pannenberg, *Jesus: God and Man*, 112.

[26] J. Moltmann, *The Crucified God* (London: SCM Press, 1974), 162.

[27] Lapide, *The Resurrection of Jesus*, 89.

[28] D. Jenkins, *The Glory of Man* (London: SCM Press, 1967), 89.

intuition that in the end all will be well. Max Horkheimer spoke of the wistful longing that the murderer should not triumph over his innocent victim. The resurrection of Jesus is the sign that such human hope is not delusory. It is the antidote to that human dread of the threat of non-being on which the existentialist tradition from Kierkegaard onwards has laid such emphasis. This is so because it is part of Christian understanding that what happened to Jesus within history is a foretaste and guarantee of what will await all of us beyond history. 'For as in Adam all die, so also in Christ shall all be made alive' (1 Cor. 15.22); 'he is the beginning, the first-born from the dead, that in everything he might be pre-eminent' (Col. 1.18).

The resurrection of Jesus is a great act of God, but its singularity is its timing, not its nature, for it is a historical anticipation of the eschato-logical destiny of the whole of humankind. The resurrection is the begin-ning of God's great act of redemptive transformation, the seed from which the new creation begins to grow (cf. 2 Cor. 5.17). Pannenberg says, 'Only the *eschaton* will ultimately disclose what really happened in Jesus' resurrection from the dead.'[29] Rahner says that 'By his resurrection and ascension Jesus did not merely enter into a pre-existent heaven; rather his resurrection created heaven for us.'[30] When John Robinson says that 'The "new thing" that God is doing is always concerned with the re-creation of the old rather than with its scrapping and supercession – He takes up the continuities to remake them,'[31] one thinks immediately of the continuity-in-discontinuity of the resurrected Jesus, the glorified body which bears the scars of the passion.[32]

---

[29] Pannenberg, *Jesus: God and Man*, 367.

[30] Quoted in Macquarrie, *Jesus in Modern Thought*, 410.

[31] J. A. T. Robinson, *Redating the New Testament* (London: SCM Press, 1976), 49.

[32] The preceding section is from *The Faith of a Physicist* by John Polkinghorne © 1994 by John Polkinghorne, published in the USA by Princeton University Press, reprinted by permission of Princeton University Press, 119–22.

19

Trinitarian theology

There are five topics that seem particularly relevant to a Trinitarian engagement with science and religion. The first of these is:

(1) God in relation to creatures

Christian theology has always strongly resisted a pantheistic identification of God and nature, of the kind that the philosophy of Benedict Spinoza endorsed. In doing so, theology has run counter to an inclination that is to be found in the thinking of a number of scientists who have sought to add a religious gloss to their feeling of wonder at the deep rational order and fruitful history of the universe. Einstein rejected belief in a personal God but he often liked to speak about 'the Old One,' employing the term as a cipher for the intellectually satisfying fundamental patterns of the physical universe that induced in him a true feeling of awe. He said that if he had a God, it was indeed the God of Spinoza, a thinker whom he greatly admired and concerning whom he once wrote a poem that begins with the line 'How much do I love that noble man.'[1]

This attitude will not do for Christianity. Its God is not a World Principle, embodied in the cosmos and so both coming into being with the origin of the universe and also fading away into nothingness when that universe eventually draws to its dying close. The Christian God is the Ground of the hope of a destiny beyond death, both for human individuals and for the cosmos itself. This thought alone requires that Christian theology make a sharp distinction between creation and its Creator, whose purposes extend beyond the ultimately futile history of the present world.

This insight is further supported by human worshipful encounter with the sacred. The numinous element in that experience testifies to the presence of the divine Other, the One who stands over against humanity in mercy and in judgement. Yet there is also a complementary element in that experience, witnessed to in the most intense way by the testimony of the mystics of all ages and traditions, which speaks of the closeness of the

---

[1] M. Jammer, *Einstein and Religion* (Princeton: Princeton University Press, 1999), 43.

divine to the human worshipper. Paul, in his 'university sermon' in the Athenian Areopagus, spoke of the God 'in whom we live and move and have our being' (Acts 17.28). Theologians need to be able to speak both of divine transcendence and of divine immanence, experienced in these ways.

It is widely recognised by many today that the theological way of thinking which may be called 'classical theism,' expressed in the West by the tradition running from Augustine to Aquinas and on through the thought of people like Calvin, laid too great a stress on divine transcendence. Its picture of a God wholly outside created time, acting on creatures but not at all acted on by them, so distanced the Creator from creation as to seem to imperil the fundamental Christian conviction of the love of God for that creation. A great deal of recent theological thinking has sought to redress the balance between divine transcendence and divine immanence.

Among scientist-theologians, a popular way in which to seek to do so has been supposed to lie in panentheism, the belief that 'the Being of God includes and penetrates the whole universe, so that every part of it exists in Him, but His Being is more than, and not exhausted by, the universe.'[2] The language employed is not free from ambiguity. The word 'penetrates' need imply no more than the immanent divine presence to the created universe, but the word 'includes,' placed in parallel with it, seems to point to some closer form of ontological relationship. Similar ambiguity attaches to how precisely we are to construe existing 'in Him' (which, when Paul the Jew used a similar phrase in his Athenian address, surely meant no more than an assertion of divine immanence), together with God's being described as being 'more' than the universe, which seems to imply that the universe is, in fact, part of the divine being. Arthur Peacocke has denied that panentheism treats the world as part of God, but when he writes that 'God is in all the creative processes of his creation and they are all equally "acts of God" for he is at all times present and active in them as their agent,'[3] these words seem to amount either to an endorsement of classical theology's assertion of an all-pervading divine primary causality co-present within the nexus of secondary creaturely causalities (an idea whose coherence is far from obvious), or to the incorporation of creation within the divine in some way. More recently, Peacocke has explicitly dissented from process theology's picture of equal divine participation in

---

[2] F. L. Cross and E. A. Livingstone (eds), *The Oxford Dictionary of the Christian Church*, 3rd edn (Oxford: Oxford University Press, 1997), 1213.

[3] A. R. Peacocke, *Creation and the World of Science* (Oxford: Oxford University Press, 1979), 204.

all events.[4] Of course, it is process theology that underlies Ian Barbour's acceptance of the panentheistic idea, but one must recognise that process thinking results in a highly qualified form of divine association with creation. Although the divine lure is part of every actual occasion, seeking to draw the event's outcome in a preferred direction, the particular con-crescent result that actually occurs is determined by the occasion itself, implying a considerable degree of effective detachment of creation from the God who exercises persuasion but possesses no direct power.

Recently, Philip Clayton has given a careful and extended defence of panentheism from the point of view of a philosophical theologian.[5] But even his discussion is not free from an acknowledged degree of semantic plasticity, as the occasional use of phrases like 'in a sense' indicates.[6] From a philosophical point of view, one of the considerations held to point to panentheism is the need for the divine infinity to be absolutely inclusive. Clayton says that 'it turns out to be impossible to conceive of God as fully infinite if he is limited by something outside himself.'[7] No doubt that would be true if the limitation were really externally imposed upon God, but what if the limitation is one that is freely internally accepted by the divine Love as the necessary cost of holding in being a creation that has been endowed by its Creator with the freedom that allows it to be itself and to make itself? A very important contemporary theological insight is the recognition that the act of creation is an act of divine kenosis, pre-cisely involving a self-limitation of this sort.[8] It involves a kind of divine 'making way' for the existence of the created other. Kenosis is the fulfilment of God's power, not its curtailment, for it is the expression of the Creator's love for creation.

This kenotic insight is of particular significance in relation to questions of theodicy, for it implies that not everything that happens – neither the act of a murderer nor the incidence of a cancer – is brought about dir-ectly by God or is in accordance with the divine will. Here is a problem to which panentheistic thinkers seem to have paid too little attention. The more closely God is identified with creation, the more acute become the problems posed by the existence of evil within that creation.

---

[4] A. R. Peacocke, *Theology for a Scientific Age*, enlarged edition (London: SCM Press, 1993), 372.

[5] P. D. Clayton, *God and Contemporary Science* (Edinburgh: Edinburgh University Press, 1997), esp. ch. 4. For a critique, see J. C. Polkinghorne, *Faith, Science and Understanding* (SPCK/Yale University Press, 2000), § 5.3.

[6] Ibid., 90, 94, 99, 100; see also 102; 'we are composed, metaphorically speaking at least, out of God.'

[7] Ibid., 99.

[8] See J. C. Polkinghorne (ed.), *The Work of Love* (SPCK/Eerdmans, 2001).

The classic Christian doctrine of the distinction between Creator and creation has come under question before in the course of the Church's history. In earlier centuries the threat came from neo-Platonism's alternative view, which depicted the world as being an emanation from the divine, existing at the outermost and attenuated fringe of deity. Panentheism represents a kind of modern version of emanationism, in its strong emphasis on the need for an absolute divine inclusivity. It is, of course, extremely difficult for anyone to be totally coherent and consistent in their thinking about so profound a matter as the relation of the infinite Creator to finite creatures. Everyone is in danger of trying to impose too easily some form of logical grid upon an inherently mysterious matter. Our commonsense notions derived from everyday experience cannot be expected to serve adequately for thinking about the profundity of God's relationship to humanity or creation generally. Not to acknowledge this would be to end up in a position similar to that of a classical physicist who refused to recognise the idiosyncrasy of the quantum world. Neither in science nor in theology may we expect to escape entirely from paradoxical tensions, but we should only embrace paradox and mystery when they are forced upon us by the sheer undeniability of the character of what it is that is being experienced.

I certainly believe that the distant God of classical theism, existing in isolated transcendence, is a concept in need of correction by a recovered recognition of the immanent presence of the Creator to creation. However, I do not believe that this requires us to embrace the too-inclusive language of panentheism. All that is necessary is to reaffirm that creatures live in the divine presence and in the context of the activity of the living God. A concept that seems to be of value here is the distinction made in the thinking of the Eastern Church, and particularly in the writings of Maximus the Confessor and Gregory Palamas, between the divine essence (God's being, ineffable to creatures) and the divine energies (God's activity in creation). An appropriate understanding of the latter can provide a strong account of effective divine presence without endangering the distinction between creatures and their Creator.[9] I wish to consider the energies as immanently active divine operations *ad extra*.

---

[9] Orthodox thinking preserves a strict distinction between the Uncreated and the created, but some forms of its discourse on the going forth of the uncreated divine energies may seem to be in danger of verging on a form of emanationism (see the complex and nuanced discussion in V. Lossky, *The Mystical Theology of the Eastern Church* (London: James Clarke, 1957), ch. 4). I think this tendency can arise from failing sufficiently to emphasise that *theosis* is a process that requires eschatological completion (see J. C. Polkinghorne, *The God of Hope and the End of the World* (SPCK/Yale University Press, 2002), 115 and 132–6).

Acts of providence are to be understood in accordance with a recognition of the divine kenosis involved in creation, so that God is not supposed to be the agent of everything but, rather, a balance is struck between the actions of God and the actions of creatures. I have suggested, despite much theological argument to the contrary, that the Creator's self-limitation should be understood to extend even to God's condescending to act as a providential cause among causes.[10]

(2) Trinitarian thinking

The first Christians were monotheistic Jews who knew above all else that the God of Israel is one Lord. Yet they found that in writing and preaching about their experience of the risen and exalted Christ, they were driven to use divine language about him, even to the point of granting him the title 'Lord,' which was the peculiar prerogative of the God of Israel. They also knew a divine Spirit at work in their hearts and lives, which sometimes they called the Spirit of God, sometimes the Spirit of Christ and sometimes just the (Holy) Spirit. The New Testament leaves these tensions and paradoxes unresolved, but obviously matters could not be allowed to remain for long in this intellectually unstable state. After more than three centuries of intense theological reflection and struggle, the Church formulated and embraced the doctrine of the Trinity, expressing its belief that the one true God exists in the eternal interchange of love between the three divine Persons, Father, Son and Holy Spirit.

It is important to me that Trinitarian thinking arose primarily as a response to the insistent complexity of human encounter with the reality of God experienced within the growing life of the Church, and not as an act of unbridled and ungrounded metaphysical speculation. It is congenial to the thinking habits of a scientist to approach the doctrine of the Trinity 'from below' and to understand it as derived from the experience of salvation. This approach locates the theological basis of Trinitarian thinking in what the Greeks called the 'economy' (*oikonomia*), the knowledge of God that arises from the Creator's interaction with creatures. However, Catherine LaCugna, in her survey of Trinitarian theology, warns us against placing too much reliance on an approach from below:

> Theology cannot be reduced to soteriology. Nor can trinitarian theology be
> purely functional; trinitarian theology is not merely a summary of our

---

[10] J. C. Polkinghorne, 'Kenotic Creation and Divine Action', in Polkinghorne (ed.), *The Work of Love*, 104–5.

experience of God. It is this, but it is also a statement, however partial, about the mystery of God's eternal being.[11]

Of course, I agree that Trinitarian thinking is not *merely* a summary of experience, any more than science is *merely* a positivistic summary of experimental data. But it is from that data that science gets its nudge in the direction of a deeper and more comprehensive understanding. Similarly, it is from the Church's experience, both soteriological and doxological, that it gets its nudge in the direction of the doctrine of the Holy Trinity. In thinking about theology I am always very conscious of the question, so natural to the scientist, 'What makes you think that this might be the case?'

The rooting of understanding in experience is what I have called bottom-up thinking. Accordingly, the approach to Trinitarian thought that I find most helpful and persuasive is one that follows the strategy expressed in the celebrated theological aphorism called 'Rahner's Rule,'[12] affirming the identity of the immanent Trinity (God in Godhead itself) with the economic Trinity (God known through creation and salvation). In other words, I rely on the belief that God's nature is truly made known through God's revelatory acts. Rahner's Rule seems to me a statement of theological realism, the assertion that what we know is a trustworthy guide to the way things are. In the case of theology, this trust is directly underwritten by the faithfulness of the God so revealed, who will not be a deceiver. It was in this spirit that I wrote that 'The proclamation of the One in Three and the Three in One is not a piece of mystical arithmetic, but a summary of data.'[13]

On reflection, it would have been better to have written 'interpretation' rather than 'summary' for, just as scientific theories are not simply read out from nature but require creative understanding of nature, so, to an even higher degree because of the veiled character of God's presence, Trinitarian thinking demands the use of creative, and indeed inspired, insight in the handling of its data. In the course of this process, second-order reflection may lead to the modification of relatively naive first-order categorisations of that experience. An example of this happening in theology is provided by the arguments that took place in the early Church

---

[11] C. LaCugna, *God For Us* (San Francisco: HarperOne, 1993), 4.

[12] K. Rahner, *The Trinity* (London: Burns & Oates, 1970), 22.

[13] J. C. Polkinghorne, *Science and Christian Belief/The Faith of a Physicist* (SPCK/Princeton University Press, 1994), 154.

about the relationships of the Persons to each other, and the degree, if any, of subordination that might characterise these relationships.

If one simply takes the testimony of the economy at face value, a subordinationist account seems to be the natural conclusion. In John's gospel, Jesus says 'The Father is greater than I' (John 14.28) and there is a repeated emphasis on the *sending* of the Son by the Father, a theme also to be found in Paul (Romans 8.3). All four gospels portray Jesus as praying to his Heavenly Father. Arianism's subordination of the Son to the Father drew its support from just these types of scriptural passage. LaCugna comments that pre-Nicene Christian thinking in general concentrated primarily on the economy and this led to what she characterises as a 'patently subordinationist christology.'[14] (One sees this reflected in Irenaeus's description,[15] of the Second and Third Persons as the 'hands of God.') However, further theological reflection after the Council of Nicaea led the Church to the recognition of the unsatisfactoriness of a position that simply and naively equated the necessary historical phenomenology of the incarnation with the eternal divine realities.

An approach from below will always have to reckon with the possibility that its understanding will need to rise above simple appropriation of the phenomena that lie at its base. (In science this corresponds to the difference between phenomenology and fundamental theory.) The essential requirement is that when theology expands and modifies its preliminary understanding, it does so for well-motivated reasons and not just in an excess of speculative exuberance.

Persuasive theoretical ideas find their support, over and above the initiating evidence, in a number of ways. One is the additional insight that can be provided, going beyond the considerations that led to the proposal being made in the first place. This overplus of interpretative success is one of the characteristics of deep scientific theories and it is an important factor in encouraging the belief that such theories do indeed afford us verisimilitudinous accounts of the actual nature of the physical world.[16] Trinitarian theology does not lack a similar kind of support. An unqualified monotheistic picture of God can only interpret the statement 'God is love' (1 John 4.8) as implying that from all eternity there has been within the divine nature itself an almost narcissistic state of self-regard.

---

[14] LaCugna, *God For Us*, 23.

[15] E. Osborn, *Irenaeus of Lyons* (Cambridge: Cambridge University Press, 2001), 89–93.

[16] See J. C. Polkinghorne, *Beyond Science* (Cambridge: Cambridge University Press, 1996), ch. 2.

The Trinitarian picture of the eternal exchange of love between the divine Persons, whose communion of mutual openness constitutes the divine being, is a much more illuminating theological insight. No doubt, its articulation requires great subtlety, as theologians seek to avoid making the Trinity so 'social' that it becomes more or less a tritheistic pantheon. The concepts of perichoresis and appropriation, baffling as they may sometimes seem to the bottom-up thinker, clearly are intended to meet this need to avoid tritheism. It is beyond my limited abilities to be sure exactly how successfully this aim has been achieved by the theological proposals that have been made.

Another way in which second-order theoretical proposals can be justified is by their gaining collateral support from their consistency with other aspects of human knowledge lying outside the field of motivation for the original ideas. I suggest that the profound degree of relationality that science has found to be present in the fabric of the physical universe is certainly congenial to a Trinitarian way of thinking.[17]

---

[17] The preceding section is from *Science and the Trinity* by John Polkinghorne © 2004 by Yale University, published in the USA by Yale University Press, reprinted by permission of Yale University Press, 93–103.

20

Eucharist

The bottom-up thinker seeks to move from experience to understanding. In science this transition occurs under the pressure of the questions that the experimentalists pose to the theorists, often pushing them to the discovery of ideas that they would not have been able to find, or even to imagine, without the necessary nudge of nature. J. B. S. Haldane once said that the world is not only stranger than we thought, it is stranger than we could have thought. One might say that the reasoning of science advances by a process of universe-assisted logic.

I believe that there is an analogous movement in theology, where the faith that is seeking understanding receives its impetus from religious experience.[1] This experience is of a variety of kinds. A substantial part, of course, is vicarious, deriving from the acceptance of the accounts of the foundational events and insights recorded in scripture. A further part also comes to us externally, from the testimony of outstanding religious figures, the kind of 'pattern-setters' that so interested William James in his important Gifford Lectures, *The Varieties of Religious Experience.*[2] Every theologian must make use of these resources, which together constitute a canonical body of classical Christian literature.[3] But, for a living encounter between faith and understanding, there must also be an internal resource. The ancient meaning of the word 'theologian' was not narrowly academic, but referred to someone who not only thought about God but was also a person of prayer, someone whose own religious experience was an indispensable part of their theological life. I wish to number myself as a humble member of that company, though I must confess that I am not a case that would have been of any interest to William

---

[1] See J. C. Polkinghorne, *Reason and Reality* (SPCK/Trinity Press International, 1991), ch. 1; *Belief in God in an Age of Science* (Yale University Press, 1998), 116–22. Defence of the evidential value of religious experience is given in R. Swinburne, *The Existence of God* (Oxford: Oxford University Press, 1979), ch. 13; W. P. Alston, *Perceiving God* (Ithaca/London: Cornell University Press, 1991).

[2] W. James, *The Varieties of Religious Experience* (London: Longmans, Green & Co., 1902).

[3] See D. Tracy, *The Analogical Imagination* (London: SCM Press, 1981).

James. My Christian life is central to who I am, but I have to acknowledge that mine is a rather humdrum kind of spirituality. For me, Christian practice centres on a certain degree of faithfulness in prayer, worship and service. A particularly important part of this experience is located in my regular participation, week by week, in the Eucharistic celebration of the Church. Sometimes I have the privilege of presiding at the altar on behalf of the gathered community of the faithful, and sometimes I am simply a member of the congregation. Whatever the role, that regular sharing in Holy Communion is an indispensable element in my Christian life. For me, theological thinking proceeds by a kind of 'liturgy-assisted logic.'[4] I want to explore a little of what that might mean. Science as such will be relevant to this task only to the extent that it encourages reliance on interpreted experience as the route to truth.

The first thing that I want to say is that I understand this life of regular Eucharistic practice to be a fulfilment of the Lord's command to do this in remembrance (*anamnesis*) of him. Of course, I know that to make this assertion immediately raises tricky scholarly questions.[5] The oldest account we have of the institution of the Lord's Supper is given by Paul in his first letter to the Corinthians (11.23–26), which conveys the dominical command for remembrance in relation to both the bread and the cup. In Luke (22.19–20), the explicit command is given in relation to the bread only. At least that is so in all the early Greek manuscripts except for Codex Bezae, which gives a severely truncated account, omitting the command altogether. In spite of the weight of manuscript testimony supporting the longer reading, many scholars have favoured attributing authenticity to the shorter text, principally on the grounds of the text-critical injunction generally to prefer the shorter reading, since it is easier to imagine how material might subsequently be added than to imagine how it would come to have been omitted. In Mark (14.22–24), and in the essentially parallel passage in Matthew (26.26–28), there is no command to make remembrance. All the accounts have Jesus associating the bread with his body and the cup with his blood, the latter linked in some way with covenant.

The assessment of this material is complicated by the fact that the 'breaking of bread' is testified to have been a regular practice of the Christian community from the very first (Acts 2.46) and so the way in which the words of Jesus were recalled and conveyed to others would have been subject to primitive liturgical influence and shaping from the beginning.

---

4 Polkinghorne, *Reason and Reality*, 19.
5 See J. Jeremias, *The Eucharistic Words of Jesus* (London: SCM Press, 1966).

From a theological point of view, however, I do not think one has to attain absolute certainty about historical detail in order to be able to understand the continuing celebration of the Eucharist over the ensuing centuries as being the Church's fulfillment of a command from its Lord. It seems absolutely clear to me that the Lord's Supper derives from the words and actions of Jesus himself on the night of his betrayal. One cannot suppose that the identification of the wine with the blood, so naturally repulsive an idea to normal Jewish thinking (cf. Genesis 9.4 and Acts 15.20b), could have arisen and been universally accepted in the first generation of Christians unless it had been known to have dominical authority. If this had not been the case, there would surely have been evidence of dissent in the early Church about the issue, comparable to that which did indeed arise about the status of circumcision.

When one reads Paul prefacing his account of the sacrament by saying that he received it 'from the Lord,' there are a variety of ways in which this phrase might be understood. One would be that Paul received a report handed on from those who had known Jesus in the flesh, giving a direct historical reminiscence of what he had actually said at the Last Supper. This seems to me the most probable meaning, but there are other possibilities. A much more speculative idea would be that during the period of the resurrection appearances the risen Christ conveyed more teaching to the apostles than the small amount recorded in the gospels and that the command to make remembrance was part of this 'secret teaching.' Or one could believe that the command to continue the remembrance was conveyed by the inspiration of the Holy Spirit, the Spirit of Christ, at work in the primitive Church in the immediate aftermath of Easter and Pentecost. Whichever may be the right way to think about the matter, if we believe in the continuing Trinitarian activity of God in the Church, it is indeed 'from the Lord' that we receive the command to do this in remembrance of him, and it is our duty and our joy to be obedient to that injunction.[6]

* * *

There are certain sacramental understandings that would be widely agreed. First, and of very great importance, is the recognition that Holy Communion is celebrated *in the presence of the risen Christ*. In modern liturgies, the

---

[6] The preceding section is from *Science and the Trinity* by John Polkinghorne © 2004 by Yale University, published in the USA by Yale University Press, reprinted by permission of Yale University Press, 118–22.

prayer of thanksgiving often begins with the president affirming 'The Lord is here,' to which the congregation reply 'His Spirit is with us.' Over history there have been many and bitter arguments about whether and how there is a real presence of Christ in the celebration of the Eucharist. Positions were taken ranging from a Zwinglian symbolic understanding of the sacrament, which saw it as acting simply as a sign to prompt the spiritual reception of Christ by the individual faithful believer, to a Tridentine insistence on the transubstantiated nature of the elements, which made them the carriers of the adorable and persisting reality of the body and blood of the Lord. While these differences have not totally disappeared, there is an increasing recognition that the heart of the matter lies with the reality that Jesus Christ is present in his risen and exalted personhood in the whole action of the Eucharist, the sacrament at which Christ is both the giver and the gift. Here is the core of our common Eucharistic experience, the phenomenon to which Christians wish to bear their grateful testimony, even amid a continuing and not wholly compatible variety of theological proposals for how that phenomenon is best to be understood.

It seems to me that this acknowledgement of the Lord's presence should serve as a sufficient basis for the possibility of the Eucharistic hospitality that so many of us yearn to see coming to be shared among more and more Christian people. Such a sharing certainly requires the foundation of a commonality of intent in order to sustain it, but this is surely adequately provided by the desire to meet in the presence of the risen Lord, and by the common conviction that he has promised to honour the covenant of just such a sacramental encounter. Sharing does not seem to require that all present subscribe to a particular and detailed dogmatic understanding of sacramental theology, and it would appear doubtful whether this would in fact be the case for any body of worshippers in any church community. It is entirely possible to recognise others who share one's acknowledgement of the authority of the Bible, without all being agreed about exactly how that authority is best defined. In a similar way, one can discern a Christian comradeship with those who reverently approach the sacrament affirming their expectation of a promised meeting with the real presence of the risen Christ, however they would express their understanding of that reality.

While the divine context of Holy Communion is the presence of Christ, there is also simultaneously another context, human in its nature and quite different in its character. It goes back to the Last Supper itself, for it is *the context of threat and betrayal.* This insight is not widely recognised, but its significance has recently been powerfully emphasised by Welker. He reminds us that 'Holy Communion is instituted in the night when Jesus Christ is betrayed and handed over to the powers of this world. It continually bears the impact of this background.'[7] Welker draws a contrast with the Jewish Passover meal. Its foundational context is also that of a threatened community, in this case held in slavery by the powers of Egypt. At the Passover, though the setting is one of affliction and distress, the threat of danger comes only from outside the Israelite community itself.

In the case of the Last Supper there is also external threat, stemming from the activity of the Jewish leaders and the Roman authorities (the religious and civil powers of the day), as they begin to mobilise their forces for the removal of the dangerous embarrassment, and threat to the continuing stability of society on their terms, that is posed by Jesus of Nazareth. But here there is also internal danger and betrayal. Judas, the one who will hand Jesus over to suffering and death, is sitting there at the table. The scriptural record gives us absolutely no reason to suppose that he did not also receive a piece of bread and take a sip of wine after Jesus had spoken those strange words about his body and his blood. Peter is also there, the one who in a few hours' time is to deny point-blank and with anger that he has even heard of Jesus. The other disciples are also part of the company – those who will run for their lives as soon as danger threatens. Even at the table, they were pretty jumpy, nervously saying 'Surely, not I?' (Mark 14.19) when Jesus straight out predicted that one of them would betray him.

The ambiguous character of those gathered for the Eucharist has continued in the Church. That was why Paul had to issue a stern warning to the Corinthians that 'Whoever eats the bread or drinks the cup of the Lord in an unworthy manner will be answerable for the body and blood of the Lord' (1 Corinthians 11.27). These insights place the Church in a delicate and difficult situation. The solemn nature of Holy Communion means that it cannot be an occasion on which anything goes. Yet, consideration of the company present at the Last Supper does not encourage the idea of 'fencing the table,' excluding from the sacrament anyone whose position seems to be at all dubious when subjected to scrutiny. Even less are we encouraged to accept the notion that the sacrament offers a handy means of ecclesiastical discipline, so that excommunication is a useful sanction to employ against the heretical and the recalcitrant. One cannot

---

[7] M. Welker, *What Happens in Holy Communion?* (Grand Rapids, Mich.: Eerdmans, 2000), 43.

help recalling all those dinner parties in the gospels at which Jesus is present and where the guests are questionable people like tax collectors and other sinners. It is clear that Jesus' scandalous willingness to eat with such persons, without insisting on prior public acts of repentance, gave considerable offence to his respectable religious contemporaries (Mark 2.15–17). Yet it is also impossible to believe that these meals were occasions of moral laissez-faire, which had no transforming effect on those who were participants. They were surely times when people began to be drawn out of the old life into a new and ennobled way of life, committed to following the way of Jesus. When Jesus invited himself to dinner with that old scoundrel Zacchaeus, not only did the crowd grumble but Zacchaeus's life was changed as he made fourfold restitution for his cheating and gave half of his possessions to the poor (Luke 19.1–10).

The Eucharist is too profound an occasion to be free of discipline altogether, but the reconciling and accepting nature of the sacrament is such that it must never be allowed to become the private preserve of an ecclesiastical in-group. A churchwarden is right to restrain a tipsy reveller who has wandered idly into church at the Midnight Mass on Christmas Eve, but the sacrament is not the possession of the perfect, for it is the place where sinners can find welcome and forgiveness.

One of the most important aspects of Eucharistic theology, recovered and widely acknowledged today, is the insight that considers the sacrament in its entirety, seeing it as taking place in *the total action of the gathered Christian community*. A great deal of medieval and Reformation thinking on the Eucharist was minutely focused, concentrating on a so-called 'moment of consecration' and, in consequence, centering its attention on the nature and status of the Eucharistic elements following that consecration, and especially discussing what resulted from the repetition of Christ's words of institution. Modern thinking, including much Roman Catholic thinking in the wake of Vatican II, is much more holistic in its approach. It is the whole action of the Supper that is celebrated in the presence of its risen and exalted Lord. Many important understandings follow from adopting this holistic approach.

One is a welcome relaxation of tension about precisely what happens exactly when and exactly where. The essential affirmation that needs to be made is seen to be that the Lord is present in the whole sacramental action; the body and blood of Christ are received in the course of the full unfolding of the Eucharist. The central prayer of the liturgy is indeed a prayer of thanksgiving, and not simply a formula for successful consecration. The attempt to affirm too much specificity, and to subject Holy

Communion to too detailed an analysis, has often had the result of distorting sacramental theology. In the case of Roman Catholic thinking this has tended to carry the danger of too great a degree of reification, focusing on the consecrated elements; in the case of Protestant thinking this has tended to carry the danger of too great a degree of spiritualisation, focusing on the feelings of the individual worshipper. It is the total action of the whole gathered people of God, present with the gifts that they have laid upon the holy table, that constitutes the valid sacrament. The Holy Mystery is a sacramental process through which the faithful worshipper receives the body and blood of our Lord Jesus Christ, and not a single identifiable and isolatable moment within the service.

In this Eucharistic action there is a special role for those who have been authorised by the Christian community to act on its behalf, who therefore have the privilege of presiding at the celebration of the sacrament and at the administration of communion. In my own Christian tradition, this privilege attaches to the presbyterial office, in which the ordained priest is authorised to act on behalf of the priestly company of all believers. I am grateful to be a member of that historic order, but I do not suppose that those ministers whose authorisation is conveyed to them by a different process are incapable of celebrating a valid Eucharist. Certainly, all Christian traditions must have some form of authorisation of their sacramental ministers, since the privilege conveyed is not one that can be taken up by anyone lightly, on the spur of the moment and solely on their own initiative. I also believe that none of us has the fullest possible degree of sacramental authorisation, since that could only come from the whole Church of God and our present regrettable divisions withhold that total degree of authorisation from everyone. In consequence, when two Christian bodies are reconciled to each other and fully accept each other's ministries, I feel no difficulty about there being an appropriate service involving the mutual laying on of hands as a process symbolising and effecting the fusion of two previously separated Eucharistic communities. This would not deny what each had previously received, but it would enhance the fullness of the continuing gift.

Although there will be a presiding minister at Holy Communion, acting visibly on behalf of Christ, the true host at the sacramental meal, the fact that the Eucharist is the action of the whole gathered community implies that the rest of the members of the congregation are not there in a subsidiary or spectatorial role. Their participation is as important and essential as that of the celebrant, precisely because all share in the fundamental priesthood of all believers (1 Peter 2.9–10). In my own church, canon law forbids the celebration of Communion by a priest alone.

The communal character of the sacrament means that *the communion is with each other as well as with God*. In the gospel of Matthew, Jesus bids us be reconciled with each other before we bring our gifts to the altar (5.23). There was an old tradition of Anglican piety that counselled people to go to the sacrament and to return from it without speaking to anyone. The intention was reverent but its form was distorted. Today, fortunately, this practice is largely forgotten. In fact, our problem lies in altogether the opposite direction. Often, the way that modern liturgy is celebrated can appear to encourage a kind of chatty sacramental ambience that seems in danger of taking the presence of Christ on too easy and familiar terms, laying great stress on the horizontal relationships between the worshippers at the expense of an adequate recognition of the vertical relationship with the One who is worshipped. I have been in churches in which the high point of the service has appeared to be the exchange of the Peace, resulting in prolonged excursions round the building to greet and hug friends. The practice has seemed to imply that this is what Holy Communion is really about, with the subsequent reception of the sacrament almost being something of an anticlimax.

What role then do *the gifts of bread and wine* have in all this? They are surely of great importance, but not in a manner that is detachable from the totality of what is going on. It seems to me of great significance that bread and wine are not only gifts of created nature in that they derive from wheat and grapes, but are also the products of human labour. In liturgical words that are often used at the Offertory, the gifts are 'what earth has given and human hands have made.' They represent the drawing together, in the action of the Eucharist, of the fruits of nature and the fruits of human work and skill in the total offering of creation. For that reason, Welker suggests, I am sure rightly, that the bread and the wine could never properly be replaced by purely natural products, such as water and apples.[8] The Eucharistic gifts unite nature and human culture. One is reminded of that wonderful picture in Revelation of the worship of heaven, in which the four and twenty elders (representing humanity) and the four living creatures (representing the non-human creation) fall down together in worship and thanksgiving to 'The Lord God the Almighty, who was and is and is to come' (Revelation 4.6–11). I think also of the words of Augustine, Bishop of Hippo, in one of his sermons to his congregation,

---

[8] Ibid., 66.

reminding them that they are there on the altar together with their gifts.

In these ways, the bread and wine that we receive at Communion are integrated into a profound and all-embracing context, which offers us also some insight into how it is that these gifts become for us truly the means by which we receive the body and blood of Christ. There are certainly mysteries here that are difficult to understand. I want to reaffirm my conviction stated in an earlier writing, that 'In some manner the bread and wine are an integral part of the whole Eucharistic action in a way neither detachably magical nor [simply and] dispensably symbolic.'[9] I am encouraged that the first Anglican–Roman Catholic International Commission, commenting on the statement that the bread and wine become the body and blood of Christ, said,

> *Becoming* does not imply material change. Nor does the liturgical use of the word imply that the bread and wine become Christ's body and blood in such a way that in the eucharistic celebration his presence is limited to the consecrated elements. It does not imply that Christ becomes present in the eucharist in the same manner that he was present in his earthly life. It does not imply that this *becoming* follows the physical laws of this world.[10]

I understand the last point to refer to the way in which we can see the sacrament as involving some anticipation, here and now, of the final eschatological reality of God's new creation.

The linkage of community and gifts in the single action of the Eucharist is the reason why I personally am unable to share in a certain kind of extra-Eucharistic devotion to the consecrated elements. This kind of 'tabernacle piety,' centering on meditation before the reserved sacrament, has been subject to some re-evaluation within the Roman Catholic community in recent years. Father Bouyer comments that, in its extreme forms, there had come to be a danger that 'The mass becomes merely a means to refilling the tabernacle.'[11] Of course there is no difficulty about the reservation of the sacrament so that subsequently it can be taken to the sick or the housebound, for this is the way in which they participate with the other worshippers in a single extended Eucharistic celebration.

Sacramental experience is very rich and many-layered, and there is always a danger in any age that the temper of the times will encourage

---

[9] J. C. Polkinghorne, *Science and Providence* (SPCK, 1989), 94.

[10] ARCIC – *The Final Report* (London: CTS/SPCK, 1982), 21.

[11] L. Bouyer, *Eucharist: Theology and Spirituality and the Eucharistic Prayer* (Notre Dame, Ind.: University of Notre Dame Press, 1989), 10.

the neglect or distortion of some important aspect of its character. One might suspect that this has happened today in a lack of attention to a dimension of the Holy Communion that was, perhaps, of obsessive concern in Reformation times. I refer to the understanding of the Eucharist as being the place where we receive *the forgiveness of sins and liberation from the power of sin*. We certainly need to recall the gospel insight that, according to Matthew, on the night of his betrayal Jesus spoke of his 'blood of the covenant which is poured out for many for the forgiveness of sins' (Matthew 26.28). Welker detects a degree of tension in Eucharistic theology between this insight and the complementary insight, strongly emphasised in the Johannine teaching, that looks to the dimension of eschatological hope expressed in the words of Jesus, 'Those who eat my flesh and drink my blood have eternal life and I will raise them up at the last day' (John 6.54). Welker regards a concern with sins forgiven as more characteristic of Reformation sacramental theology and a concern with being sustained in eternal life as more characteristic of Catholic thought. He links the theme of liberation from the power of sin with the significance that he attributes to the context of compromise and threat in which the sacrament was instituted and in which it continues to be celebrated.[12]

The acknowledgement of human need for deliverance from sin is important and it can act as an antidote to too easy and unthinking an approach to the Holy Mysteries. Yet, many Reformation liturgies expressed an obsessive preoccupation with the issue of sin and its remedy, creating imbalance that reflected the particular salvific concerns of the time, which centred on how it could be that sinful human beings received acceptance from a holy God. The Anglican Book of Common Prayer of 1662, drawing much of its material from the work of Thomas Cranmer more than a century earlier, is a good example of this tendency. Its Communion liturgy has a heavy emphasis on penitential material and on the proferring of reassurance to those who are conscious of their past failures. Confession and absolution are immediately followed by the further reassurance of the 'Comfortable Words,' which in turn are followed by a further Prayer of Humble Access, before the worshipper is considered to be ready to approach the Communion table. Theologically, the focus is almost entirely on redemption through the cross of Christ. One of the major driving forces in producing modern liturgies within Anglicanism has been so that they can offer a better balance between cross and resurrection in their

---

[12] Welker, *What Happens in Holy Communion*, ch. 10.

articulation of the gospel, and so express both the forgiveness of sins and the gift of eternal life, in a way that did not prove possible for Cranmer in his response to the spiritual atmosphere of the sixteenth century. One of the gifts that we have received from the Reformers, for which we should be grateful, is the recognition that the benefits that flow from Christ are not given us quasi-magically or mechanically, *ex opere operato*, but have to be received fittingly by the faithful as they participate in the total action of the Lord's Supper.

Further important dimensions of the Eucharist demand our attention. The first of these is that, though the Eucharist is celebrated on each occasion at a particular time and in a particular place, nevertheless it has a universal character that brings into focus all places and *the plenitude of times, past, present and still to come*. In many modern liturgies, the whole congregation, in the course of the great prayer of thanksgiving, is called on to affirm that 'Christ has died, Christ is risen, Christ will come again.' The Eucharist is celebrated now, in the presence of the risen Christ who is ever contemporary, while at the same time it looks back to the events of Jesus' earthly life, in particular to his sacrificial death on Calvary, and it looks forward expectantly to the hope of the eschatological future, already beginning to come into being through the seminal event of Christ's resurrection from the dead, and eventually to be fulfilled by the visible vindication of his Lordship, of which the Second Coming is the symbol.

Paul told the Corinthians that 'as often as you eat the bread and drink the cup, you proclaim the Lord's death until he come' (1 Corinthians 11.26). The aspect of the Eucharist that relates to time past focuses particularly on the cross. The broken bread recalls the Lord's broken body and the poured-out wine recalls his shed blood. These 'remembrances of him' are something very much more powerful than simple historical reminiscence. The Greek word translated 'memory' is *anamnesis*, and it can carry the force of an event of the past that is re-presented now, for contemporary participation in it, as when Jewish people at Passover still share in the Exodus event of all those centuries ago. Similarly, the Christian worshipper at Holy Communion participates in the sacrificial event of Christ's death on the cross. In the twenty-first century, sacrificial language is difficult for many people, but it has always played an important role in sacramental theology. Today there is virtually universal recognition that the Mass is not a re-enactment or simple continuation of Christ's once-for-all death on our behalf at Calvary, but in the Eucharist there is an anamnetic re-presentation of that sacrifice, enabling us to participate truly in that sacrificial event, as we offer ourselves to the Father through the

Son and receive the benefits of Christ's passion through the Holy Spirit's work within us, in the doxological context of worshipful praise.[13]

* * *

Our exploration of Eucharistic theology has illustrated to some extent the depth and subtlety of the kind of interpreted experience that constitutes the motivation for Christian beliefs. Sacramental theology is as complex and sophisticated, and ultimately as powerfully insightful, as the considerations that support a fundamental theory in science. In neither case would one expect the lines of argument to be superficially simple or naively accessible. In each case, the cost of illumination is the willingness to have one's everyday habits of thought revised and expanded under the influence of the reality encountered.[14]

---

[13] The preceding section is from *Science and the Trinity* by John Polkinghorne © 2004 by Yale University, published in the USA by Yale University Press, reprinted by permission of Yale University Press, 126–38.

[14] The preceding section is from *Science and the Trinity* by John Polkinghorne © 2004 by Yale University, published in the USA by Yale University Press, reprinted by permission of Yale University Press, 141.

21

Eschatology

As we begin our consideration of the theological approach to eschato-logical issues, we need to return yet again to the question of hope. This theological virtue is a matter of central concern for a credible articulation of Christian belief, which must seek a total understanding of God and God's purposes, capable of embracing not only the possibilities of the present but also the sufferings of the past and the expectations of the future. Jürgen Moltmann said that 'From first to last, and not merely in the epilogue, Christianity is eschatology, is hope, forward looking and forward moving, and therefore also revolutionary and transforming the present.'[1] Fifty years earlier, Karl Barth had said more or less the same: 'Christianity that is not entirely and altogether eschatology has entirely and altogether nothing to do with Christ.'[2]

Hope is the negation both of Promethean presumption, which sup-poses that fulfilment is always potentially there, ready for human grasping, and also of despair, which supposes that there will never be fulfilment, but only a succession of broken dreams. Hope is quite distinct also from a utopian myth of progress, which privileges the future over the past, seeing the ills and frustrations of earlier generations as being no more than necessary stepping stones to better things in prospect.

If eschatology is to make sense, all the generations of history must attain their ultimate and individual meaning. Christianity takes the reality of evil seriously, with all the perplexities that entails. It 'refuses the premature consolation that pre-empts grief, the facile optimism which cannot recognize evil for what it is.'[3] As part of its unflinching engagement with reality, Christianity will recognise the seriousness of science's prediction of ultimate cosmic futility. As part of its unflinching engagement with history, Christianity will recognize that episodes like the Holocaust deny

---

[1] J. Moltmann, *Theology of Hope* (London: SCM Press, 1967), 16.

[2] K. Barth, *The Epistle to the Romans* (Oxford: Oxford University Press, 1933), 314.

[3] R. Bauckham and T. Hart, *Hope Against Hope* (London: Darton, Longman & Todd, 1999), 42.

to it any shallow conception of what hope for the future might mean, as if it could be divorced from acknowledgement of the horror of the past.

Holding in mind such a clear-eyed view of the woes and disappointments of history, one must ask what could then be the ground of a true hope beyond history? There is only one possible source: the eternal faithfulness of the God who is the Creator and Redeemer of history. Here Christianity relies heavily upon its Jewish roots. It is only God who can bring new life and raise the dead, whose Spirit breathes life into dry bones and makes them live (Ezekiel 37.9–10). Hope lies in the divine *chesed*, God's steadfast love, and not in some Hellenistic belief in an unchanging realm of ideas or an intrinsic immortality of the human soul. Christian trust in divine faithfulness is reinforced by the knowledge that God is the One who raised Jesus from the dead. Only such a God could be the ground for that hope against hope that transcends the limits of any natural expectation.

This means that a credible eschatology must find its basis in a 'thick' and developed theology. A kind of minimalist account of deity, which sees God as not much more than the Mind behind cosmic order, will not be adequate. Nor will a kind of minimalist Christology, which sees Jesus as no more than an inspired teacher, pointing humanity to new possibilities for self-realisation and with his message living on in the minds of his followers, provide a sufficient insight into the divine purposes for creation beyond its death to be the ground of an everlasting hope. These concepts are too weak to bear so great a weight of expectation. To sustain true hope it must be possible to speak of a God who is powerful and active, not simply holding creation in being but also interacting with its history, the one who 'gives life to the dead and calls into existence the things that do not exist' (Romans 4.17). This same God must be the one whose loving concern for individual creatures is such that the divine power will be brought into play to bring about these creatures' everlasting good. The God and Father of our Lord Jesus Christ is just such a God. To be persuasive, eschatological hope requires more than a general intuition that something must survive death. The problems that beset the realistic hope of a post mortem destiny are complex and demanding. They call for a corresponding richness and depth in our understanding of the power and steadfast love of God.

The question of eschatological hope is also the question of the fundamental meaningfulness of human life within creation. Are those moments of our deep experience when we glimpse that reality is trustworthy and that all will be well, intimations of our ultimate destiny or merely fleeting and illusory consolations in a world of actual and absolute transience? Moltmann says, 'Our question about life, consequently, is not whether our existence might possibly be immortal, and if so what part of it; the question is: *will love endure*, the love out of which we receive ourselves, and which makes us living when we ourselves offer it.'[4] If God is, as Christians believe, the God of love, then love will indeed endure. 'Many waters cannot quench love, neither can floods drown it' (Song of Solomon 8.7) – not even the waters of cosmic chaos nor the tumultuous breakers of human evil.

Forgiveness and joy

Hope, then, must involve the redemption of the past as well as a promised fulfilment in the future. Indeed, the one requires the other. If it is to be true and total, hope must look in both directions. One may ask where participation in such an all-embracing hope could find its setting in human life. Two important sources are our experiences of forgiveness and of joy, the one freeing us from the tyranny of the past, the other offering us a foretaste of the ultimate future.

Without forgiveness there can be no redemption of the past. God's forgiveness comes to us through the cross of Christ, 'the lamb of God who takes away the sin of the world' (John 1.29). This forgiveness frees us from the shackles with which we have enslaved ourselves. Equally necessary for our liberation is the forgiveness that we give to others for the hurts that they have inflicted on us. Resentment and the desire to pay back are distorting and corrupting influences from which we must seek to be released. In the New Testament, the receiving and giving of forgiveness are often seen as parts of a single action, linked together because they are mutually necessary (the Lord's prayer: Matthew 6.12; 6.14–15; 18.35; and so on).

It is a costly business to forgive a real wrong, and a costly business also to receive forgiveness for a real wrong committed. Such actions centre on the recognition of a painful reality. They are far removed from the trivial indifference of 'It doesn't matter.' The drunken motorist actually killed the innocent child and he has to acknowledge that this is so. In an act of astonishing generosity, the bereaved parent can nevertheless rise above the natural desire to seek punitive revenge. The grace of God is powerfully at

---

[4] J. Moltmann, *The Coming of God* (London: SCM Press, 1996), 53; my italics.

work in such a situation. Acts of forgiveness offer experience of a ground of hope for the redemption and healing of the past. Christoph Schwöbel writes, 'Every experience of gratuitous forgiveness offers vindication of eschatological hope.'[5]

Our experiences of joy, those deep moments of peaceful happiness that come to us through music, art, nature and human love, and through the worship of God, are foretastes of the fruits of eschatological fulfilment (cf. 2 Corinthians 1.22). These are insights of a dynamic and unifying kind. Miroslav Volf says that 'Joy lives from *the movement in time* qualified by an unperturbed peace between past and future in all presents.'[6] A vision of what that might ultimately mean is beautifully expressed in a prayer based on some words of John Donne:

> Bring us, O Lord, at our last awakening into the house and gate of heav'n, to enter into that gate and dwell in that house, where there shall be no darkness nor dazzling, but one equal light; no noise nor silence, but one equal music; no fears nor hopes, but one equal possession; no ends nor beginnings, but one equal eternity; in the habitation of thy glory and dominion, world without end.

Realised eschatology

But is it necessary for there to be a 'last awakening' for this kind of experience to be consummated? In *The End of the World and the Ends of God*, Kathryn Tanner presented a powerful account of a realised eschatology of the present moment, whose ultimate quality lies precisely in its character of being this life lived in relationship with God. In her 'thought experiment' she argued for an *eschatologia continua*, paralleling the widely accepted concept of *creatio continua* in which God's unfolding creative purposes are being fulfilled within the evolutionary processes of the universe. Just as the emphasis of the doctrine of creation properly lies in God's upholding of the world in being at every moment and not in the instant of its beginning, so Tanner believes that the emphasis of eschatology should lie in the attainment of a life lived with God now and not in some future state of blessedness beyond death. 'Old and new are found

---

[5] C. Schwöbel, 'The Church as a Cultural Space', in John Polkinghorne and Michael Welker (eds), *The End of the World and the Ends of God* (Chicago: Trinity Press, 2000), 122.

[6] M. Volf, 'Enter into Joy! Sin, Death, and the Life of the World to Come', in John Polkinghorne and Michael Welker (eds), *The End of the World and the Ends of God* (Chicago: Trinity Press, 2000), 275.

together in the world we know.'[7] She comments that 'so understood, eternal life presents a more spatialised than temporalised eschatology.'[8] It is life lived at the present moment in the presence of God. Such ideas have been popular with many theologians, from Friedrich Schleiermacher to Paul Tillich.

Yet, while an emphasis of this kind offers a healthy corrective to a purely futuristic, 'pie in the sky' kind of eschatology, it seems by itself to be an inadequate expression of the Christian hope. This life is too hurtful and incomplete to be the whole story. What are we to make of those whose lives are tragically cut short, or grievously oppressed and distorted by their circumstances? They too must have their share in the kingdom of God. Tanner is strong in making a call for action to right the wrongs of this world: 'complacency is ruled out not by a transcendent future but by a transcendent present by the present life in God as the source of the goods that the world one lives in fails to match.'[9] But such action can, at best, be only half the story. Without a transcendent future, many are condemned to a loss of good that no process solely within history could ever restore to them. In fact, all of us are so condemned, even if we have the good fortune to die in honoured and pious old age. We shall all die with unfinished business and incompleteness in our lives. There must be more to hope for.

Similar difficulties attend the concept of an atemporal 'objective immortality' that simply sees each human life as being preserved in its totality in the eternal memory of God, an idea that has been popular with process theologians and with some other thinkers.[10] Either such lives are preserved as much with their sins as with their good deeds, as much with their frustrations as with their achievements, or they are held in a purged and purified form that would be a false memory of the one to whom it purports to relate. Actual eschatological fulfilment demands for each of us a completion that can be attained only if we have a continuing and developing personal relationship with God post mortem. We must participate in our own salvation. As Miroslav Volf says, speaking of God's ultimate act in the redemptive justification of the sinner, 'In soteriology, the "objects" of justification are always persons, never their done deeds or lived lives.'[11]

---

[7] K. Tanner, 'Eschatology without a Future?', in John Polkinghorne and Michael Welker (eds), *The End of the World and the Ends of God* (Chicago: Trinity Press, 2000), 236.

[8] Ibid., 230.

[9] Ibid., 234.

[10] See J. Hick, *Death and Eternal Life* (London: Collins, 1976), 104–9; Volf, 'Enter into Joy!', 259–70.

[11] Volf, 'Enter into Joy!', 263.

The basic issue here is whether temporality is constitutive of being truly human, an essential good and not an unfortunate deficiency.[12]

* * *

If human beings have a destiny beyond death that is much more than a mere resuscitation (which would amount to no more than another turn of the wheel of this transient world), then what is it that will connect our present life to our future life in that new world whose character will be so different? We face here, in a critical way, the issue of continuity and discontinuity that lies at the heart of all attempts to gain a degree of eschatological understanding. What could it be that ensures that it is indeed Abraham, Isaac and Jacob who live in the kingdom of God, and not just new beings who have been given the old names?

The soul

In the course of Christian thinking, an answer to these questions has frequently been made by appeal to the concept of the soul, conceived of in a platonic fashion as a spiritual entity, released from imprisonment in the fleshly body at the moment of death. While there are still body/soul dualists of this kind,[13] for many people this has become an extremely problematic way of conceiving of human nature. Our evolutionary history appears to link us in a continuous way with our primate ancestry, which in turn can be traced back through simpler life forms to the bacteria who, for 2 billion years, were the sole living inhabitants of Earth. Although it cannot absolutely be ruled out that at some stage, the Creator adjoined a separate and additional spiritual component to complement evolving bodies of increasing complexity, once those bodies had reached the appropriate stage of development, the idea seems contrived and unpersuasive to many. They find greater theological satisfaction in the concept of the divine sustaining of a process of continuous creation through evolutionary development.

The striking effects that physical incidents, such as brain damage or drug intake, can have on human personality also encourage taking a psychosomatic view of the nature of human beings. A celebrated case, often referred to in this connection, is that of Phineas Gage. In 1848, this efficient and

---

[12] The preceding section is from *The God of Hope and the End of the World* by John Polkinghorne © 2002 by Yale University, published in the USA by Yale University Press, reprinted by permission of Yale University Press, 93–100.

[13] J. C. Eccles, *The Human Mystery* (Berlin/New York: Springer, 1979); R. Swinburne, *The Evolution of the Soul* (Oxford: Oxford University Press, 1986).

capable construction foreman was involved in a terrible accident in which a premature explosion drove an iron bar through the front of his brain and out of the top of his head. Astonishingly, he survived and within two months was declared physically healed. However, his personality was completely changed. Gage had become fitful and capricious, endlessly restive and quite incapable of holding down a job. It was clear that the gross damage to the brain that he had sustained had completely changed the character and equilibrium of his mind. We may conclude, from this and much other evidence, that human beings look much more like animated bodies than like incarnated souls.

Despite both Hebrew thought and much of the thinking of the New Testament being in accord with this unitary view of human nature, a difficulty might be feared to ensue for the coherence of eschatological hope. We must ask whether, in consequence of this anti-dualist conclusion, one has lost the possibility of speaking of the human soul altogether, so that there is no longer a way in which we may frame an understanding of a destiny beyond death, expressed in terms of the soul's provision of the necessary element of continuity required to make such a belief meaningful. I do not think that this is the case.[14]

Whatever the human soul may be, it is surely what expresses and carries the continuity of living personhood. We already face within this life the problem of what that entity might be. The soul must be the 'real me' that links the boy of childhood to the ageing academic of later life. If that carrier of continuity is not a separate spiritual component, what else could it be? It is certainly not merely material. The atoms that make up our bodies are continuously being replaced in the course of wear and tear, eating and drinking. We have very few atoms in our bodies today that were there even two years ago. What does appear to be the carrier of continuity is the immensely complex 'information-bearing pattern' in which that matter is organised. This pattern is not static; it is modified as we acquire new experiences, insights and memories, in accordance with the dynamic of our living history. It is this information-bearing pattern that is the soul.[15]

---

[14] J. C. Polkinghorne, *Science and Christian Belief/The Faith of a Physicist* (SPCK/Princeton University Press, 1994), 163.

[15] The point of view presented here corresponds to a form of dual-aspect monism (see J. C. Polkinghorne, *Faith, Science and Understanding* (SPCK/Yale University Press, 2000), ch. 5.4). It bears some relationship to the non-reductive physicalism of W. Brown, N. Murphy and H. N. Malony (eds), *Whatever Happened to the Human Soul?* (Philadelphia: Fortress, 1999). However, I prefer the more evenly balanced dual-aspect terminology, and I do not set store by supervenience as an explanatory category.

Modern science, through its study of complex systems, is beginning to recognize the importance of information as a complement to energy in the description of the process of the world. The concept of the soul that has just been proposed is fully in accord with this development. I believe that we can follow Thomas Aquinas in adopting, in appropriately modern phrasing and understanding, the concept of the soul as the form, or information-bearing pattern, of the body. Of course, our present understanding of these profound matters is not sufficient for us to be able to frame an adequate account of what such a conception could mean in detail, but there does seem to be the prospect of a coherent, if inevitably somewhat conjectural, way of holding together human psychosomatic unity with human personal identity. It would be altogether too crude to say that the soul is the software running on the hardware of the body – for we have good reason to believe that human beings are very much more than 'computers made of meat'[16] – but that unsatisfactory image catches a little of what is being proposed.

While the soul, understood in this way, has a dynamic and changing character, it is perfectly possible to suppose that, amid its evolving change, each individual soul carries specific elements of its patterning which are the signature of its own abiding and unique personal identity. (A mathematician would say that there were invariant characters, preserved in the course of unfolding transformation.)

Finally, we may note that it would even be possible to reconcile this concept of the soul with a highly modified form of platonism. We have spoken earlier of an everlasting realm of mathematical entities. It would be conceivable that the information-bearing patterns of the soul could be considered as intersecting with this realm and that they would remain lodged there after the decay of the body. However, for the reasons already given, I prefer a thorough-going psychosomatic picture of human nature, in which the preservation of the soul depends only on divine faithfulness. Of course, the resurrection re-embodiment of the soul would in any case have to be God's act.

Destiny beyond death

If these ideas contain some truth, we have to acknowledge that this information-bearing pattern will, in the course of nature, be dissolved by

---

[16] R. Penrose, *The Emperor's New Mind* (Oxford: Oxford University Press, 1989), ch. 10; J. Searle, *Minds, Brains and Science* (London: BBC Publications, 1984).

the decay of our bodies after death. There is, therefore, no intrinsic immortality associated with the soul in this way of understanding it. Death is a real end. However, it need not be an ultimate end, for in Christian understanding only God is ultimate. It is a perfectly coherent hope that the pattern that is a human being could be held in the divine memory after that person's death. Such a disembodied existence, even if located within the divine remembrance, would be less than fully human. It would be more like the Hebrew concept of shades in Sheol, though now a Sheol from which the Lord was not absent but, quite to the contrary, God was sustaining it. It is a further coherent hope, and one for which the resurrection of Jesus Christ provides the foretaste and guarantee, that God in the eschatological future will re-embody this multitude of preserved information-bearing patterns in some new environment of God's choosing.

In other words, there is indeed the Christian hope of a destiny beyond death, but it resides not in the presumed immortality of a spiritual soul, but in the divinely guaranteed eschatological sequence of death and resurrection. Only a hope conceived of in this way can do full justice to human psychosomatic unity, and hence to the indispensability of some form of re-embodiment for a truly human future existence. The only ground for this hope – and the sufficient ground for this hope, as we have already emphasised – lies in the faithfulness of the Creator, in the unrelenting divine love for all creatures.

Some philosophers have objected to the idea of re-embodiment without intervening physical continuity, on the grounds that if it were possible, then what would prevent the multiplication of replicas, with the incoherence of personal identity that would result. The answer is surely that only God has the power to effect such re-embodiment and divine consistency would never permit the duplication of a person.

A number of further comments need to be made. As expressed so far, the emphasis placed on information-bearing pattern has had a strongly individualistic tone. However, we must recognise that the deep relationality of creation, and the significant distinction between a human person (constituted in relationships) and a mere individual (treated as if existing in self-isolation), encourage a broader view. The 'pattern that is me' cannot adequately be expressed without its having a collective dimension. In this connection, it is significant that a powerful way of articulating Christian eschatological destiny is through the incorporation of believers into the one 'body of Christ' (1 Corinthians 12.27; Ephesians 4.12–13). It is also important to recognise, as Miroslav Volf emphasises, that eschatological fulfilment must involve the mutual reconciliation of human beings. 'Persons

217

cannot be healed without the healing of their specific socially constructed and temporarily [*sic*] constructed identities.'[17]

An immediate, intimate and pastorally sensitive issue is how we are to conceive of the relationship between the living and the dead. Moltmann presses upon us the importance of addressing the question, 'Where are the dead?'[18] Earlier, he had written that 'the idea of an enduring communion between the living and the dead in Christ, and of the community of Christ as a communion of the living and the dead is a good and necessary one.'[19] Moltmann recalls that in the base communities of South America, when the roll is called of the names of those who 'disappeared' in the troubles, the congregation all say 'Presente.' He comments that 'the community of the living and the dead is the praxis of resurrection hope.'[20]

If the souls awaiting the final resurrection are held in the mind of God, as we have suggested, then 'in the Lord' there will surely be a mediated relationship between the living and the dead. One of the most natural ways in which to express this relationship will be through prayer. No doubt, those who are 'in Christ' are wholly within God's loving care and protection, but we should not argue that this makes it unnecessary or inappropriate for us to pray for them. After all, the same is as true of those among the living who are committed to Christ, as it is of the departed. Many arguments alleged against praying for the dead seem, on the face of it, to apply equally to praying for the living. Of course, we must rid our prayers for the dead of some unfortunate and unedifying medieval distortions. We are not involved in an instrumental manipulation on their behalf, as an unreformed notion of 'masses for the dead' might have seemed to suggest. Prayer is always mutual participation in grace and never the exercise of a quasi-magical power. It is significant that the Eastern Orthodox Christians, always so sensitive to eschatological reality, hold intercessions for the dead but offer no masses on their behalf.

In the patristic period there was some speculation about what age people will be at the resurrection. A popular answer was about thirty; not only because it could be seen as corresponding to some climactic in human life but also because it was believed to be the age at which Jesus died and rose again. In terms of our present discussion, this ancient argument could

---

[17] Volf, 'Enter into Joy!', 262.
[18] J. Moltmann, 'Is there Life After Death?', in John Polkinghorne and Michael Welker (eds), *The End of the World and the Ends of God* (Chicago: Trinity Press, 2000), 246–7.
[19] Moltmann, *The Coming of God*, 98.
[20] Ibid., 108.

be rephrased by asking at what state of its dynamic development will the information-bearing pattern of the human person be reinstantiated in the re-embodied resurrection life. As with many other detailed eschatological questionings, one might be tempted to reply, 'Wait and see.' Yet a very serious concern lies behind the query, not least in relation to those who through enduring dementia and the ravages of Alzheimer's disease have suffered a kind of partial death within the confines of this life. A similar issue relates to those who through severe congenital disability may be thought, in some ways at least, only to have attained to a limited kind of life in this world. We might ask, also, in what condition will Phineas Gage be resurrected?

We do not need to suppose that being held in the mind of God is a purely passive kind of preservation. We may expect that God's love will be at work, through the respectful but powerful operation of divine grace, purifying and transforming the souls awaiting resurrection in ways that respect their integrity. Ultimately, what has been lost will be restored and what of good was never gained will be bestowed. We do not need to suppose that the divine eschatological process of redemptive fulfilment is wholly confined to the resurrected life of the new creation. It may begin in whatever post mortem state precedes that final destiny, a thought that the New Testament language of 'Paradise' (Luke 23.43) and being 'with Christ' (Philippians 1.23) may be held to encourage. Yet its ultimate fulfilment must wait upon the restoration of complete embodied human personality. The workings of divine grace will not only involve the healing of disability and the restoration of decay, but it may also be expected to begin its work within all of us, for we shall all need God's sanctifying and redeeming touch beyond what we have already experienced in this world.[21]

* * *

God must surely care for all creatures in ways that accord with their natures. Therefore, we must expect that there will be a destiny for the whole universe beyond its death, just as there will be a post mortem destiny for humankind. We have seen that two remarkable New Testament passages (Romans 8.18–25; Colossians 1.15–20) do indeed speak of a cosmic

---

[21] The preceding section is from *The God of Hope and the End of the World* by John Polkinghorne © 2002 by Yale University, published in the USA by Yale University Press, reprinted by permission of Yale University Press, 103–12.

redemption. Just as we see Jesus' resurrection as the origin and guarantee of human hope, so we can also see it as the origin and guarantee of a universal hope. The significance of the empty tomb is that the Lord's risen and glorified body is the transmuted form of his dead body. Thus matter itself participates in the resurrection transformation, enjoying thereby the foretaste of its own redemption from decay. The resurrection of Jesus is the seminal event from which the whole of God's new creation has already begun to grow.

The risen Christ is no resuscitated corpse. His body has new properties that enable it, at his will, to appear and disappear within present history. Nor will the redeemed universe be a mere repetition of its present state. This current universe is a creation endowed with just those physical properties that have enabled it to 'make itself' in the course of its evolving history. A world of this kind, by its necessary nature, must be a world of transience in which death is the cost of new life. In theological terms, this world is a creation that is sustained by its Creator, and which has been endowed with a divinely purposed fruitfulness, but which is also allowed to be at some distance from the veiled presence of the One who holds it in being and interacts in hidden ways with its history. Its unfolding process develops within the 'space' that God has given it, within which it is allowed to be itself. This is a theme that has been developed particularly by Jürgen Moltmann.[22] He draws on the Kabbalistic notion of *zimzum*, the divine making way for the existence of created reality. One may sum up this insight by saying that this creation is the result of a kenotic act by the Creator, who has made way for the existence of the created other.[23] The physical fabric of such a universe must take a particular form, but there is no reason to suppose that the Creator cannot bring into being a new creation of a different character when it is appropriate to the divine purpose to do so.

The world to come will indeed have to have a different character. Just as Jesus was exalted to the right hand of the Father after his resurrection, so the world to come will be integrated in a new and intimate way with the divine life. I do not accept panentheism (the idea that the creation is in God, though God exceeds creation)[24] as a theological reality for the present world, but I do believe in it as the form of eschatological destiny

---

[22] See J. Moltmann, *The Trinity and the Kingdom of God* (London: SCM Press, 1981), 105–13; *God in Creation* (London: SCM Press, 1985), ch. 4.

[23] See J. C. Polkinghorne (ed.), *The Work of Love* (SPCK/Eerdmans, 2001).

[24] See Polkinghorne, *Faith, Science and Understanding*, ch. 5.3.

for the world to come. As Paul wrote to the Corinthians, God will then be 'all in all' (1 Corinthians 15.28). The Eastern Orthodox speak of eschatological fulfilment as being the attainment of *theosis*, not meaning by that that creatures will become gods but that they will share fully in the divine life and energies. This world is one that contains the focussed and covenanted occasions of divine presence that we call sacraments. The new creation will be wholly sacramental, suffused with the presence of the life of God. In his great vision of the End, the seer of Patmos saw the holy city as one in which there was no longer a cultic temple 'for its temple is the Lord God the Almighty and the Lamb' (Revelation 21.22). God's presence, veiled from us today, will be open and manifest in the world to come. Moltmann has his own way of expressing this hope, in terms of the descent of the divine Shekinah.

These are great hopes, in which the necessarily discontinuous side of eschatological expectation finds its expression. Much has to be taken on trust, for it is clearly beyond our feeble powers to conceive exactly how a redeemed universe will function. Yet it seems a coherent hope to believe that the laws of its nature will be perfectly adapted to the everlasting life of that world where 'Death will be no more; mourning and crying and pain will be no more, for the first things have passed away' (Revelation 21.4), just as the laws of nature of this world are perfectly adapted to the character of its freely evolving process, through which the old creation has made itself.

The equally necessary continuity between the old and new creations lies in the fact that the latter is the redeemed transform of the former. The pattern for this is the resurrection of Christ where, as we have already emphasised, the Lord's risen body is the eschatological transform of his dead body. This implies that the new creation does not arise from a radically novel creative act *ex nihilo*, but as a redemptive act *ex vetere*, out of the old.[25]

Important theological consequences flow from this understanding. The pressing question of why the Creator brought into being this vale of tears if it is the case that God can eventually create a world that is free from suffering, here finds its answer. God's total creative intent is seen to be intrinsically a two-step process: first the old creation, allowed to explore and realise its potentiality at some metaphysical distance from its Creator; then the redeemed new creation which, through the Cosmic Christ, is

---

[25] Polkinghorne, *Science and Christian Belief/Faith of a Physicist*, 167.

brought into a freely embraced and intimate relationship with the life of God.

A further consequence of this understanding is that it clearly establishes the value of the old creation, since it affords the raw material for eschatological transformation into the new creation. An other-worldly negation of a duty of environmental care for this present world is thereby made impossible. I find this to be a firmer and more realistic basis for the affirmation of the worth of the present world than that which might be provided by dubious utopian or millenarian speculations.

The world to come

It is, of course, the transmuted 'matter' of the new creation that will be the setting for human re-embodiment in the resurrection life. Paul was emphatic that 'flesh and blood cannot inherit the kingdom of God, nor does the perishable inherit the imperishable' (1 Corinthians 15.50). Much of the fifteenth chapter of 1 Corinthians is devoted to Paul's wrestling with what this discontinuity within continuity might mean, making use of the science of his day (notably the image of the seed that 'dies'; 1 Corinthians 15.35–41). We might try the same endeavour with the aid of contemporary scientific resources, but I think that we will fare no better in terms of any attempted detailed discussion of the nature of the new creation. One thing, however, we may reasonably anticipate.

In this universe, space, time and matter are all mutually interlinked in the single package deal of general relativity. It seems reasonable to suppose that this linkage is a general feature of the Creator's will. If so, the new creation will also have its 'space' and 'time' and 'matter.' The most significant theological consequence of this belief is the expectation that there will be 'time' in the world to come.[26]

* * *

Speaking of the new as arising from the transformation of the old might seem to encourage a simple notion of their successive relationship, so that future 'time' is the continuation of present time beyond its natural end. Yet the occurrence of the resurrection of Christ as an event within, as well

---

[26] The preceding section is from *The God of Hope and the End of the World* by John Polkinghorne © 2002 by Yale University, published in the USA by Yale University Press, reprinted by permission of Yale University Press, 113–17.

as beyond, present history, suggests the necessity for a more nuanced notion of the connection between old and new. It is possible, and theologically desirable, to consider a subtler relationship rather than simple succession. Mathematicians can readily think of the spacetime of the old creation and the 'spacetime' of the new creation as being in different dimensions of the totality of divinely sustained reality, with resurrection involving an information-bearing mapping between the two, and the redemption of matter as involving a projection from the old onto the new. Such a picture offers some partial insight into the nature of the appearances of the risen Christ, as arising from limited intersections between these two worlds. It also offers a possible response to the observation that, if the new creation simply follows on from the old, God will have to wait an awful long time if our universe is, in fact going to expand forever, so that its expected future will be an infinitely prolonged dying fall.[27]

If something like this is indeed the case, it offers us ways of thinking about how life in this world and life in the world to come might be related to each other. Much traditional Christian thinking about an intermediate state between death and resurrection has been in terms of 'soul sleep,' a kind of suspended animation awaiting the restoration of full humanity. Our idea of the information-bearing patterns of souls being held in the mind of God has some obvious kinship with this picture. Some modern theologians, including Karl Rahner, have thought about the matter differently, supposing that, though we all die at different times in this world, we may all arrive simultaneously on the day of resurrection in the world to come. If time and 'time' are related in the way we have been discussing, this would clearly be a coherent possibility, with mappings from different times all leading to the same 'time' in the world to come. This would constitute the modern version of what we might suppose Paul to have been expressing when he spoke of us all being changed 'in the twinkling of an eye' as the perishable puts on imperishability (1 Corinthians 15.52–53). It might also help us with the understanding of the meaning of a verse, much discussed by writers on eschatology, in which Jesus says to the penitent thief, 'Truly I tell you, today you will be with me in Paradise' (Luke 23.43).

---

[27] Other possible responses would include either God's bringing the dying fall to an end once it had gone beyond a certain level of diminished cosmic activity at which creativity had effectively ceased, or to query whether divine experience of created temporality is measured on human scientific scales (see Psalm 90.4).

All creation

One further topic remains to be discussed in relation to the new creation. What are we to expect will be the destiny of non-human creatures? They must have their share in cosmic hope, but we scarcely need suppose that every dinosaur that ever lived, let alone all of the vast multitude of bacteria that have constituted so large a fraction of biomass throughout the history of terrestrial life, will each have its own individual eschatological future. On the other hand, the kind of theological thinking that has too exclusively an anthropocentric focus surely takes too narrow a view of God's creative purposes. It is conceivable that this eschatological dilemma can be resolved by according significance in non-human creatures more to the type than to the token. Many people who are respectful of animals would nevertheless consider it permissible to cull individuals in order to preserve the herd. Perhaps there will be lions in the world to come but not every lion that has ever lived. If that is the case, lionhood will have also to share in the dialectic of eschatological continuity and discontinuity, in accordance with the prophet vision that in 'the new heavens and the new earth . . . the wolf and the lamb shall feed together, the lion shall eat straw like the ox' (Isaiah 65.17 and 25). Particularly interesting, in a speculative way, is the question of the destiny of pets, who could be thought to have acquired enhanced individual status through their interactions with humans. Perhaps they will have a particular role to play in the restored relationships of the world to come.

Some kind of balance between transience and preservation is certainly necessary. In his integrative eschatology,[28] Moltmann rightly insists that individual destinies and universal destinies are opposite sides of the same eschatological coin. Yet, when he writes that 'resurrection has become the universal "law" of creation, not merely for human beings, but for animals, plants, stones and all cosmic systems as well,'[29] the eschaton is in danger of becoming a museum collection of all that has ever been. It is hard to believe that individual stones as such either have or need an ultimate destiny.[30]

---

[28] Moltmann, *The Coming of God*, xiv–xvi.

[29] J. Moltmann, *The Way of Jesus Christ* (London: SCM Press, 1990), 258.

[30] The preceding section is from *The God of Hope and the End of the World* by John Polkinghorne © 2002 by Yale University, published in the USA by Yale University Press, reprinted by permission of Yale University Press, 120–3.

22

World faiths

The discussion of the interrelationships of the world faith traditions seeks to survey a truly ecumenical scene. It can only do so here in the broadest and most simplified terms. A sketch of the critical issues is the most that can be attempted. When one considers the slow rate of progress of ecumenical discussion within the Christian community of divided churches, it is clear that centuries of encounter are likely to be necessary before the world faiths make substantive progress in mutual dialogue.

The sacred

Not all the traditions are even theistic, for Theravada Buddhism is at most agnostic, with the concept of nirvana said to play something like the fundamental role played by God in other faiths. Yet there is a common concern with a realm of spiritual significance and experience, however differently that realm is construed and described, that seems to link all the traditions in a common meeting with the sacred (understood as reality transcending the immediate and mundane). The similarity of reports of mystical encounters, in whatever tradition they occur, has already been noted and to a lesser extent the same is true of encounters with the numinous. When dedicated and serious followers of different faith traditions meet each other, they can often recognize the authenticity of the spiritual experiences of the other, despite the differences of linguistic expression that may be involved in their articulation.

The world faith traditions may be understood as preserving a testimony against a reductive, materialist account of reality. They hold out the prospect of some kind of fulfilment which is to be found in the spiritual realm. When one enquires more specifically about the character of that fulfilment, however, perplexing differences of description then become apparent.

Dissonance

Comparisons are complicated by the considerable differences present within each tradition itself, and also by the differing cultural settings in which each tradition finds its classic expression. Simplistic comparisons, such as the assertion that the Abrahamic faiths are world-affirming and the Eastern faiths world-denying, though crudely expressing actual differences of emphasis, run up against the tradition of the *Bhodisattvas* in Buddhism (enlightened ones who postpone their entry into nirvana in order to help others still in this world to find enlightenment in their turn) or the Desert Fathers in Christianity (whose extreme ascetism led to the rejection of the ordinary life of humankind). Cultural differences make the translation of the scriptures of the faiths an extremely delicate and difficult task, thereby increasing the problems of attaining mutual understanding. When all these caveats have been entered, it still seems possible to identify major points of disagreement between the traditions.

The human self

The Abrahamic faiths agree in assigning to individual human beings the highest significance in God's sight. Ancient Israel had a strong sense of the collective nature of the family and tribe (fathers and sons together, in the patriarchical language of the culture), but with the prophets of the Exile, such as Ezekiel, came the recognition of individual identity and responsibility and, later still, the expectation of a personal everlasting destiny beyond death. In the Gospels, Jesus is said to have asserted that such is God's care for the individual that the very hairs of our heads are numbered (Matt. 10.30).

The Eastern religions, on the other hand, and particularly Buddhism, see the self as ultimately an illusion from which to seek release, for clinging to individuality is the source of suffering. In contrast to Christianity's search for the purification that leads to right desire (the longing of the soul for God which is fundamental to the thought of Augustine), Buddhism's aim is the enlightenment that leads to non-desire.

The nature of time

Closely connected with the foregoing, there are differences in the attitudes to time that characterize the traditions. The Abrahamic faiths all have a strongly linear understanding of time, as a path to be trodden by the individual pilgrim. The Eastern faiths, on the other hand, see time in more circularly recurrent terms. This relates to the doctrine of reincarnation – so natural, it seems, to the Eastern mind, so perplexing to the Western. Those clinging to the illusion of self are destined to live a succession of lives as the wheel of *samsara* (reincarnation) revolves, until they find eventual release from this perpetual return.

Suffering

Within samsaric cyclicity, one's fate at the next turn of the wheel is determined by karma, the entail of good and evil carried forward from the past. This concept provides a ready-made, if unverifiable, explanation of the suffering to be endured in the present: it is the working out of bad karma, acquired in a previous life. For the Eastern religions, the problem of suffering is eventually solved by *anatta* (non-self) and the release from *samsara* that this brings.

For the Abrahamic faiths, suffering is a deadly reality and by no means the consequence of bondage to illusion. From the book of Job onwards, they have had to wrestle with the deep perplexities that this brings. A little of the Christian response has already been sketched, where insight is focused on the event of the cross of Christ.

History

A religion centred on the attainment of enlightenment is a religion whose principal concern is with timeless truths. It can sit light to history. Gandhi greatly valued Christ's teaching, such as in the Sermon on the Mount, but he said it was a matter of indifference to him whether Jesus had actually lived or not. Christianity is committed to the historical specificity of the life, death and resurrection of Jesus Christ, just as Judaism is committed to God's Passover deliverance out of Egypt and Islam to the life of Mohammed and the communication to him of the Qur'an. The Abrahamic faiths share a serious engagement with the reality of history, corresponding to their linear, progressive, understanding of the nature of time, within which their foundational revelatory events took place.

Monism

The Abrahamic faiths strongly emphasize the distinction between the Creator and creation. The universe is not divine. The Eastern traditions are monistic in their emphasis upon an ultimate unity of all reality, including the divine. In the advaitic tradition of Hindu thought, the Ultimate is *nirguna Brahman*, without any qualities, though religious practice makes the concession of speaking of *saguna Brahman*, reality with qualities. Monism underlines the assertion of the ultimate illusion of the self, for all are drops in the one ocean of being. This is an area in which the existence of different cultural and philosophical traditions, with the varieties of linguistic expression that go with them, enhances the danger of misunderstandings between the faiths, but it does seem that a fundamental dissonance is present.

Finally, one should note that, despite their commonalities, there are also serious dissonances within the family of Abrahamic faiths, not least in their assessments of Jesus of Nazareth. Islam regards him as one of the prophets, second only to Mohammed, and a number of contemporary Jews would wish to recognize him as an outstanding Jewish figure, though not the Messiah. Neither faith could accept the Christian belief in his divine status.

Responses

A variety of responses have been made to the clashing accounts of their encounters with the sacred, given by the world faiths.

Exclusivism

The classic Christian response was a simple assertion of the truth of Christianity and the error of religions that differed from it. In the early Christian centuries, the gods of other religions (other than the God of Israel, of course) were considered to be deceiving demons. In the nineteenth century, the same stance led an English missionary bishop in India to write a hymn speaking of how 'the heathen in his blindness bows down to wood and stone.' A principal motivation for this attitude arose from Christianity's claim of the unique and final character of God's self-revelation in Christ. In the often quoted words of Jesus in the fourth Gospel, 'No one comes to the Father except through me' (John 14.6). Yet the recognition of the authentic spiritual experience clearly contained in the other world faith traditions has made such a peremptory writing of them off an increasingly difficult position to maintain. After all, the fourth Gospel also speaks of the Word as 'the light of all people' (John 1.4).

Pluralism

This is the opposite stance, placing all faith traditions on an essentially equal footing. Of course, some limits are imposed, for adherents of this position are unlikely to be as accommodating to Satanism as to Judaism. Yet the major traditions of which we have been speaking are regarded as being equally valid, if culturally distinguishable, ways of attaining self-transcendence; and their accounts of deity, or the Ultimate, are seen as different masks behind which is hidden the ineffable Real. A great difficulty with this view is its mismatch with what the adherents of the different faiths would want to say about their own understanding of the sacred. Such a bland commonality does not at all do justice to any tradition's specificities. It is very hard to believe that the dissonances noted in the previous section can be resolved in this way. The main motivation for pluralism is the belief that the Real cannot have been unknown and inaccessible for any enduring community. However, there can be other ways of attaining this desired conclusion.

Inclusivism

This is the stance favoured by many Christians who do not want to be dismissive of the spiritual experiences of their companions in other faiths, or to claim that they have nothing to learn from them, or to believe that God has left the divine nature altogether without witness at any time or place. Inclusivism, accordingly, does not deny the presence of genuine salvific experience in the different traditions, but neither does it deny the final and definitive character of God's self-revelation in Christ. Ultimately, all must indeed come to the Father through him, if he is the unique bridge between humanity and deity. But the light of the Word has also been shining in the other traditions, much as all Christians would acknowledge it to have shone in the Jewish faith. God is always and everywhere at work (through the hidden activity of the Spirit) and no community has been without some degree of true encounter with the divine. Karl Rahner was aiming at an inclusivist statement when he called the adherents of other faiths 'anonymous Christians.' Of course, they might well wish to repudiate a description that might seem to them to amount to a Christian takeover bid, but Rahner was trying to speak of comradeship rather than annexation. Like many theological positions, inclusivism is more a statement of the boundaries within which an acceptable solution may be sought, rather than the attainment of that solution. The perplexities of dissonance still remain to be resolved.

Continuing discussion

There is clearly a correlation between these attitudes to interfaith matters and the understandings of the theological enterprise. Those taking a cognitive approach will incline to exclusivism, for propositions are either true or false. Both the experiential-expressive and cultural-linguistic approaches seem hospitable to forms of pluralism and, in fact, a desire to attain some kind of ecumenical coexistence has been a factor in encouraging these ways of conceiving of theological discourse. If religion is fundamentally about inner attitudes or patterns of community living, then one must expect much cultural variety and, indeed, encourage it, for what suits one person or society will not necessarily suit another. The considerable diversities within each faith tradition are clearly accommodating human variety in this way. However, if there is also a cognitive element within religion, a concern with what is actually the case, then the problems of dissonance remain.

Inclusivism is naturally allied to a critical realist understanding of theology: acknowledging the universal presence of encounter with the sacred; seeking to understand its different modes of expression and description, while recognizing that each tradition and community must view reality from a cultural perspective; knowing that interpretation and experience intertwine, yet believing that behind it all is an actual Reality of which we may hope to gain verisimilitudinous understanding; trusting that this Reality is such that an understanding of it is, at least to some degree, attainable by humankind.

There is a growing feeling that none of the classic interfaith approaches of exclusivism, pluralism or inclusivism, in terms of their simple categorizations, are adequate to the complexity and perplexity of the meeting of the world faith traditions. This truly ecumenical dialogue is still at a very early stage. In the concluding section, we shall consider a way in which a further small advance might be made.

Science as a meeting point

Each world faith is the jealous guardian of its central tenets and its fundamental style of discourse. Even Hinduism, the most accepting of the traditions in terms of accretions from elsewhere, sets limits to its tolerance by the rejection of what it sees as intolerably exclusive claims. Jesus can be welcomed as an *avatar* (one of a number of appearances of the divine in human form) but not accepted as the only-begotten Son of God. Initial dialogue between the traditions will have to take place at their peripheries, for encounter at the centre would be too threatening.

It is possible that the consideration of the relationship between scientific and religious understandings would provide a place of meeting where meaningful issues could be raised without inducing a merely defensive response. A great deal of the material of science and religion would be suitable for reconsideration in this ecumenical way, but only through discussions between a number of adherents of the different faiths who also shared a concern for these questions arising from science. Any single author (including, of course, the present one) cannot step far outside the world of meaning within which his or her scientific and religious experience has been realized. No one can pretend to attain some magisterial vantage point from which neutral adjudication could be given. We can listen to each other, but we cannot presume to speak for each other.

Some of the questions that would be included in the agenda for such a meeting would be:

- How do we understand the nature of the physical world and our relationship to it? What is the kind of knowledge we can attain? What is the meaning of the Eastern concept of *maya*, often understood by occidentals as asserting that we live in a world subject to the play of illusion?
- What is the relationship between religious metaphysics and quantum theory's mixture of structure and flexibility and its picture of an interconnected web of events which participate in togetherness-in-separation? Do indigenous adherents of Hinduism and Buddhism detect the same resonant correspondences that some Western writers have claimed to exist between quantum theory and Eastern thought?
- How do a cosmic evolutionary history stretching over 15 billion years and a biological evolutionary history stretching over 4 billion years relate to the creation stories of the faith traditions?
- Are the deep intelligibility of the physical world, and the 'unreasonable effectiveness of mathematics' in its scientific description, signs of the cosmic presence of Mind?
- Is the anthropic fine tuning of the laws of nature in this universe a sign of the cosmic presence of Purpose?
- How do the insights of neurophysiology, psychology and the philosophy of the mind affect our understanding of the human person? Is there a coherent and stable concept of the human self?
- What is the significance of science's prediction of eventual cosmic collapse or decay?
- Does analogy with the scientific community offer any insight into the balance within the religious life between cognitive understanding, expressive commitment and a communally conducted life?
- What role does bottom-up thinking, so natural to the scientist in the way it seeks to move from evidence to understanding, have to play in the intellectual reflection upon religious claims?

It is earnestly to be hoped that a conversation of this kind will come about. It may be expected to continue for a very long time. Bottom-up thinkers would welcome the start of conversations focused on these specific issues, for not everything that needs to be done in the ecumenical encounter of the world faith traditions can be achieved through the top-down discussion of general principles.[1]

---

[1] The preceding section is from *Science and Theology* by John Polkinghorne © 1998 published in the USA by Fortress Press, reprinted by permission of Augsburg Fortress Publishers, 120–7.

Books by J. C. Polkinghorne

1966   *The Analytic S-Matrix* (Cambridge University Press); with R. J. Eden, P. V. Lanshoff and D. I. Olive.

1979   *The Particle Play* (W. H. Freeman).

1980   *Models of High Energy Processes* (Cambridge University Press).

1983   *The Way the World Is* (Triangle; Eerdmans, 1984).

1984   *The Quantum World* (Longman/Princeton University Press, 1984; Penguin, 1986).

1986   *One World* (SPCK/Princeton University Press, 1987; Templeton Foundation Press, 2007).

1988   *Science and Creation* (SPCK/New Science Library; Templeton Foundation Press, 2006).

1989   *Rochester Roundtable* (Longman/W. H. Freeman).

1989   *Science and Providence* (SPCK/New Science Library; Templeton Foundation Press, 2005).

1991   *Reason and Reality* (SPCK/Trinity Press International).

1994   *Quarks, Chaos and Christianity* (Triangle/Crossroad, 1996; second edition: SPCK/Crossroad, 2005).

1994   *Science and Christian Belief* (SPCK); also published as *The Faith of a Physicist* (Princeton University Press, 1994; Fortress, 1996).

1995   *Serious Talk* (Trinity Press International; SCM Press, 1996).

1996   *Beyond Science* (Cambridge University Press).

1996   *Scientists as Theologians* (SPCK).

1996   *Searching for Truth* (Bible Reading Fellowship/Crossroad).

1998   *Belief in God in an Age of Science* (Yale University Press).

1998   *Science and Theology* (SPCK/Fortress).

2000   *The End of the World and the Ends of God* (Trinity Press International); edited with Michael Welker.

2000   *Faith, Science and Understanding* (SPCK/Yale University Press).

2000   *Traffic in Truth* (Canterbury Press/Fortress).

2001   *Faith in the Living God* (SPCK/Fortress); with Michael Welker.

2001   *The Work of Love* (SPCK/Eerdmans); editor.

2002   *The God of Hope and the End of the World* (SPCK/Yale University Press).

2002   *Quantum Theory: A Very Short Introduction* (Oxford University Press).

2003   *The Archbishop's School of Christianity and Science* (York Courses).

2003   *Living with Hope* (SPCK/Westminster John Knox).

2004   *Science and the Trinity* (SPCK/Yale University Press).

2005   *Exploring Reality* (SPCK/Yale University Press).

Books by J. C. Polkinghorne

2007   *From Physicist to Priest: An Autobiography* (SPCK).
2007   *Quantum Physics and Theology* (SPCK/Yale University Press).
2009   *Questions of Truth* (Westminster John Knox); with Nicholas Beale.
2010   *Theology in the Context of Science* (Yale University Press).

Bibliography

Allen, D., *Christian Belief in the Modern World* (Louisville, Ky.: Westminster John Knox Press, 1989).

Alston, W. P., *Perceiving God* (Ithaca/London: Cornell University Press, 1991).

ARCIC – *The Final Report* (London: CTS/SPCK, 1982).

Augustine, *The City of God*, trans. H. Bettenson (Harmondsworth: Penguin Books, 1972).

Baelz, P., *Prayer and Providence* (London: SCM Press, 1968).

Baillie, D. M., *God Was in Christ* (London: Faber, 1956).

Baker, J., *The Foolishness of God* (London: Darton, Longman & Todd, 1970).

Barbour, I. G., *Issues in Science and Religion* (London: SCM Press, 1966).

Barbour, I. G., *Myths, Models, and Paradigms* (London: SCM Press, 1974).

Barrow, J. D. and Tipler, F. J., *The Anthropic Cosmological Principle* (Oxford: Oxford University Press, 1986).

Barth, K., *Dogmatics in Outline* (London: SCM Press, 1949).

Barth, K., *The Epistle to the Romans* (Oxford: Oxford University Press, 1933).

Bartholomew, D., *God of Chance* (London: SCM Press, 1984).

Barton, J., *People of the Book* (Atlanta: Westminster John Knox, 1988).

Bauckham, R. and Hart, T., *Hope Against Hope* (London: Darton, Longman & Todd, 1999).

Bohm, D., *Wholeness and the Implicate Order* (London: Routledge & Kegan Paul, 1980).

Bohm, D. and Hiley, B. J., *The Undivided Universe* (London: Routledge, 1993).

Bohr, N., *Atomic Physics and Human Knowledge* (London: Wiley, 1958).

Bouyer, L., *Eucharist: Theology and Spirituality and the Eucharistic Prayer* (Notre Dame, Ind.: University of Notre Dame Press, 1989).

Bowker, J., *Licensed Insanities: Religions and Belief in God in the Contemporary World* (London: Darton, Longman & Todd/Meakin and Associates, 1987).

Brooke, J. H., *Science and Religion* (Cambridge: Cambridge University Press, 1999).

Brown, W., Murphy, N. and Malony, H. N. (eds), *Whatever Happened to the Human Soul?* (Philadelphia: Fortress, 1999).

Brümmer, V., *What Are We Doing When We Pray?* (London: SCM Press, 1984).

Burrell, D., *Knowing the Unknowable God* (Notre Dame, Ind.: University of Notre Dame Press, 1986).

Caird, G. B., *The Gospel of Luke* (New York: Penguin, 1963).

Carnes, J. R., *Axiomatics and Dogmatics* (New York: Oxford University Press, 1982).

Childs, Brevard S., *Old Testament Theology in a Canonical Context* (Philadelphia: Augsburg Fortress, 1985).

Clayton, P. D., *God and Contemporary Science* (Edinburgh: Edinburgh University Press, 1997).

Coakley, S., *Powers and Submissions* (Malden, Mass.: Blackwell, 2002).

Cobb, J. B. Jr. and Griffin, D. R., *Process Theology: An Introductory Exposition* (Louisville, Ky.: Westminster Press, 1976).

Conway-Morris, S., *Life's Solution* (Cambridge: Cambridge University Press, 2003).

Cross, F. L. and Livingstone, E. A. (eds), *The Oxford Dictionary of the Christian Church*, 3rd edition (Oxford: Oxford University Press, 1997).

Cushing, J. T., *Quantum Mechanics* (Chicago: University of Chicago Press, 1994).

Davies, P. C. W., *God and the New Physics* (London: Dent, 1983).

Davies, P. C. W., *The Mind of God* (Chicago: Simon and Schuster, 1992).

Davies, P. C. W. (ed.), *The New Physics* (Cambridge: Cambridge University Press, 1989).

Dawkins, R., *The Blind Watchmaker* (London: Longman, 1986).

Dawkins, R., *River Out of Eden* (London: Weidenfeld and Nicolson, 1995).

Dennett, D., *Darwin's Dangerous Idea* (New York/London: Simon and Schuster, 1995).

d'Espagnat, B., *The Conceptual Foundations of Quantum Mechanics* (Menlo Park, Calif.: Benjamin, 1971).

d'Espagnat, B. and Whitehouse, J. C., *Reality and the Physicist* (Cambridge: Cambridge University Press, 1989).

Dodd, C. H., *The Parables of the Kingdom* (Welwyn: James Nisbet, 1961).

Dukas, H. and Hoffmann, B. (eds), *Albert Einstein: The Human Side* (Princeton: Princeton University Press, 1979).

Dunn, J. D. G., *Unity and Diversity in the New Testament* (London: SCM Press [1977], 1980).

Eccles, J. C., *The Human Mystery* (Berlin/New York: Springer, 1979).

Eddington, A. S., *Fundamental Theory* (Cambridge: Cambridge University Press, 1946).

Farmer, H. H., *The World and God: The Study of Prayer, Providence, and the Miracle of Christian Experience* (London: Nisbet, 1935).

Farrer, A. M., *Faith and Speculation* (London: A. & C. Black, 1967).

Feyerabend, P., *Against Method* (London: Verso, 1975).

Feynman, R., *The Feynman Lectures on Physics*, vol. 3 (Redwood City, Calif.: Addison-Wesley, 1965).

Ford, J. 'What is chaos that we should be mindful of it?', in P. C. W. Davies (ed.), *The New Physics* (Cambridge: Cambridge University Press, 1989).

Frye, N., *The Great Code* (London: Routledge & Kegan Paul, 1982).

Gill, R. (ed.), *Theology and Sociology* (New York: Continuum, 1987).

Gleick, J., *Chaos* (London: Heinemann, 1988).

Goodwin, B., *How the Leopard Changed its Spots* (London: Weidenfeld & Nicolson, 1994).

Green, C. (ed.), *Karl Barth* (London: Collins, 1989).

Griffin, D. R. (ed.), *Physics and the Ultimate Significance of Time* (Albany, NY: State University of New York Press, 1986).

Harvey, A. E., *Jesus and the Constraints of History* (London: Duckworth, 1982).

Hanson, N. R., *Perception and Discovery* (San Francisco: Freeman Cooper, 1969).

Hawking, S. W., *A Brief History of Time* (New York: Bantam, 1988).

Hefner, P. J., *The Human Factor* (Minneapolis: Fortress Press, 1993).

Hengel, M., *Atonement* (London: SCM Press, 1981).

Hesbert, N., *Quantum Reality* (New York: Random House, 1985).

Hick, J., *Death and Eternal Life* (London: Collins, 1976).

Hick, J., *An Interpretation of Religion: Human Responses to the Transcendent* (New York: Macmillan, 1989).

James, W., *The Varieties of Religious Experience* (London: Longmans, Green & Company, 1902).

Jammer, M., *Einstein and Religion* (Princeton: Princeton University Press, 1999).

Jantzen, G., *God's World, God's Body* (London: Darton, Longman & Todd, 1984).

Jenkins, D., *The Glory of Man* (London: SCM Press, 1967).

Jeremias, J., *The Eucharistic Words of Jesus* (London: SCM Press, 1966).

Jeremias, J., *New Testament Theology*, vol. 1 (London: SCM Press, 1971).

Kauffman, S., *The Origins of Order* (Oxford: Oxford University Press, 1993).

Kaufman, G. D., *God the Problem* (Cambridge, Mass.: Harvard University Press, 1972).

Kenny, A., *The Metaphysics of Mind* (Oxford: Oxford University Press, 1989).

Kuhn, T., *The Structure of Scientific Revolutions*, 2nd edn (Chicago: University of Chicago Press, 1970).

LaCugna, C., *God For Us* (San Francisco: HarperOne, 1993).

Lapide, P., *The Resurrection of Jesus* (London: SPCK, 1984).

Leslie, J., *Universes* (London: Routledge, 1989).

Lewis, A. E., *Between Cross and Resurrection* (Grand Rapids, Mich.: Eerdmans, 2001).

Lewis, C. S., *Miracles* (London: Geoffrey Bles, 1947).

Lonergan, B., *Method and Theology* (London: Darton, Longman & Todd, 1972).

Lossky, V., *The Mystical Theology of the Eastern Church* (London: James Clarke, 1957).

Lucas, J. R., *Freedom and Grace* (London: SPCK, 1979).

McDonald, H. D., *The God Who Responds* (Cambridge: James Clarke, 1986).

McDonald, J. I. H., *The Resurrection* (SPCK, 1989).

Mackay, A. L., *The Harvest of the Quiet Eye* (Bristol: Institute of Physics, 1977).

McMullin, E. (ed.), *Evolution and Creation* (Notre Dame, Ind.: University of Notre Dame Press, 1985).

Macquarrie, J. *Jesus in Modern Thought* (London: SCM Press, 1990).

Margenau, H., *The Nature of Physical Reality* (New York: McGraw-Hill, 1950).

Mascall, E. L., *Christian Theology and Natural Science* (London: Longman, 1956).

Mitchell, B., *The Justification of Religious Belief* (London: Macmillan, 1973).

Moltmann, J., *The Coming of God* (London: SCM Press, 1996).

Moltmann, J., *The Crucified God* (London: SCM Press, 1974).

Moltmann, J., *God in Creation*: *An Ecological Doctrine of Creation* (London: SCM Press, 1985).

Moltmann, J., 'Is there Life After Death?', in John Polkinghorne and Michael Welker, eds, *The End of the World and the Ends of God* (Chicago: Trinity Press, 2000).

Moltmann, J., *Theology of Hope* (London: SCM Press, 1967).

Moltmann, J., *The Trinity and the Kingdom of God* (London: SCM Press, 1981).

Moltmann, J., *The Way of Jesus Christ* (London: SCM Press, 1990).

Montefiore, H., *The Probability of God* (London: SCM Press, 1985).

Moule, C. D. F., *The Origin of Christology* (Cambridge: Cambridge University Press, 1977).

Munz, P., *Our Knowledge of the Growth of Knowledge* (London: Routledge & Kegan Paul, 1985).

Neill, S. C., *The Interpretation of the New Testament* (Oxford: Oxford University Press, 1964).

Newton-Smith, W. H., *The Rationality of Science* (London: Routledge & Kegan Paul, 1981).

O'Collins, G., *Jesus Risen* (Darton, Longman & Todd, 1987).

Omnes, R., *The Interpretation of Quantum Mechanics* (Princeton: Princeton University Press, 1994).

Pailin, D. A., *God and the Process of Reality* (Oxford: Oxford University Press, 1989).

Pais, A., *Subtle is the Lord* (Oxford: Oxford University Press, 1982).

Pannenberg, W., *Jesus: God and Man* (London: SCM Press, 1968).

Pascal, B., *Pensées* (New York: Penguin, 1966).

Peacocke, A. R., *Creation and the World of Science* (Oxford: Oxford University Press, 1979).

Peacocke, A. R., *God and the New Biology* (London: Dent, 1986).

Peacocke, A. R., *Theology for a Scientific Age*, enlarged edn (London: SCM Press, 1993).

Penrose, R., *The Emperor's New Mind* (Oxford: Oxford University Press, 1989).

Peters, T. (ed.), *Cosmos and Creation* (Nashville: Abingdon Press, 1989).

Peters, T. and Bennett, G., *Bridging Science and Religion* (London: SCM Press, 2002).

Pickering, A., *Constructing Quarks* (Edinburgh: Edinburgh University Press, 1984).

Polanyi, M., *Personal Knowledge* (London: Routledge & Kegan Paul, 1958).

Polkinghorne, J. C., *Belief in God in an Age of Science* (New Haven/London: Yale University Press, 1998).

Polkinghorne, J. C., *Beyond Science* (Cambridge: Cambridge University Press, 1996).

Polkinghorne, J. C., *Faith, Science, and Understanding* (London: SPCK/Yale University Press, 2000).

Polkinghorne, J. C., *The God of Hope and the End of the World* (London: SPCK/Yale University Press, 2002).

Polkinghorne, J. C., *The Particle Play* (New York: W. H. Freeman, 1979).

Polkinghorne, J. C., *Quantum Theory: A Very Short Introduction* (Oxford: Oxford University Press, 2002).

Polkinghorne, J. C., *The Quantum World* (London/Princeton: Longman/Princeton University Press, 1984).

Polkinghorne, J. C., *Reason and Reality* (London: SPCK/Trinity Press International, 1991).

Polkinghorne, J. C., *Science and Christian Belief/The Faith of a Physicist* (London/Princeton: SPCK/Princeton University Press, 1994).

Polkinghorne, J. C., *Science and Providence* (London/Boston, Mass.: SPCK/New Science Library, 1989).

Polkinghorne, J. C., *Scientists as Theologians* (London: SPCK, 1996).

Polkinghorne, J. C. (ed.), *The Work of Love* (London: SPCK/Eerdmans, 2001).

Pollard, W. G., *Chance and Providence* (London: Faber & Faber, 1958).

Popper, K., *Conjectures and Refutations* (New York: Basic Books, 1962).

Popper, K., *The Logic of Scientific Discovery* (London: Hutchinson, 1968).

Prigogine, I., *The End of Certainty* (New York: The Free Press, 1996).

Prigogine, I., 'Time, Chaos, and the Laws of Physics' (lecture given in London, May 1995).

Prigogine, I. and Stengers, I., *Order Out of Chaos* (London: Heinemann, 1984).

Rae, A., *Quantum Physics: Illusion or Reality?* (Cambridge: Cambridge University Press, 1986).

Rahner, K., *The Trinity* (London: Burns & Oates, 1970).

Ramsey, I., *Religion and Science: Conflict and Synthesis* (London: SPCK, 1964).

Ramsey, I., *Religious Language* (London: SCM Press, 1957).

Robinson, J. A. T., *Redating the New Testament* (London: SCM Press, 1976).

Rolston, H., *Genes, Genesis, and God* (Cambridge: Cambridge University Press, 1998).

Ruse, M., *Can a Darwinian Be a Christian?* (Cambridge: Cambridge University Press, 2001).

Russell, R. J., Murphy, N. and Isham C. J. (eds), *Quantum Cosmology and the Laws of Nature* (Vatican: Vatican Observatory, 1993).

Russell, R. J., Murphy, N. and Peacocke, A. R. (eds), *Chaos and Complexity* (Vatican: Vatican Observatory, 1995).

Sanders, E. P., *Jesus and Judaism* (London: SCM Press, 1985).

Sanders, E. P., *Paul and Palestinian Judaism* (London: SCM Press, 1977).

Sanders, E. P. and Davies, M., *Studying the Synoptic Gospels* (London: SCM Press, 1989).

Schillebeeckx, E., *Jesus* (London: Collins, 1974).

Schwöbel, C., 'The Church as a Cultural Space', in John Polkinghorne and Michael Welker, eds, *The End of the World and the Ends of God* (Chicago: Trinity Press, 2000).

Searle, J., *Minds, Brains and Science* (London: BBC Publications, 1984).

Sherwin-White, A. N., *Roman Society and Roman Law in the New Testament* (Oxford/Grand Rapids, Mich.: Oxford University Press/Baker, 1963).

Soskice, J. M., *Metaphor and Religious Language* (Oxford: Clarendon Press, 1985).

Stannard, R., *Science and the Renewal of Belief* (London: SCM Press, 1982).

Stump, E. and Murray, M. J. (eds), *Philosophy of Religion: The Big Question* (Blackwell, 1999).

Swinburne, R., *The Concept of Miracle* (Basingstoke: Macmillan, 1970).

Swinburne, R., *The Evolution of the Soul* (Oxford: Oxford University Press, 1986).

Swinburne, R., *The Existence of God* (Oxford: Oxford University Press, 1979).

Swinburne, R., *Faith and Reason* (Oxford: Oxford University Press, 1981).

Tanner, K., 'Eschatology without a Future?', in John Polkinghorne and Michael Welker, eds, *The End of the World and the Ends of God* (Chicago: Trinity Press, 2000).

Torrance, T., *Theological Science* (Oxford: Oxford University Press, 1969).

Tracy, D., *The Analogical Imagination* (London: SCM Press, 1981).

van Huysteen, W., *Duet or Duel?* (London: SCM Press, 1998).

Vanstone, W. H., *Love's Endeavour, Love's Expense* (London: Darton, Longman & Todd, 1977).

Vermes, G., *Jesus the Jew* (London: SCM Press, 1983).

Volf, M., 'Enter into Joy! Sin, Death, and the Life of the World to Come', in John Polkinghorne and Michael Welker, eds, *The End of the World and the Ends of God* (Chicago: Trinity Press, 2000).

Ward, K., *The Battle for the Soul* (London: Hodder & Stoughton, 1985).

Ward, K., *Divine Action* (London: Collins, 1990).

Ward, K., *Rational Theology and the Creativity of God* (Malden, Mass.: Blackwell, 1982).

Weinberg, S., *The First Three Minutes: A Modern View of the Origin of the Universe* (New York: Basic Books, 1993).

Welker, M., *What Happens in Holy Communion?* (Grand Rapids, Mich.: Eerdmans, 2000).

White, V., *The Fall of a Sparrow* (Exeter: Paternoster Press, 1985).

Whitehead, A. N., *Process and Reality* (New York: Free Press, 1978).

Wiles, M., *God's Action in the World* (London: SCM Press, 1986).

Wilson, E. O., *Consilience* (New York: Knopf, 1998).

Wilson, E. O., *Sociobiology* (Cambridge, Mass.: Belknap Press, 1975).

Wright, N. T., *The New Testament and the People of God* (London: SPCK, 1992).

Wuketits, F., *Evolutionary Epistemology* (New York: State University of New York Press, 1990).

Index

abba 168, 169
Adam and Eve 139, 188
Aspect, Alain 32
Allen, D. 88
Alston, W. P. 197
altruism 46
Anselm 81, 101
anthropic principle 99–102, 231
Apostle Paul 5, 50, 63, 83, 139, 153,
    155, 163, 167, 177–9, 181–3, 186,
    190, 195, 198, 199, 201, 207, 221–3
Aquinas, T. 42–3, 53, 69, 72, 90, 101,
    117, 190, 216
ARCIC 205
Aristotle 42, 56, 89
atheism 59, 85, 90, 92–4, 142
Augustine 37, 57, 71, 72, 80, 91,
    108–9, 126, 131–2, 190, 204, 226
Auschwitz 71, 140

Bach, J. S. 41, 80
Baelz, P. 126
Baillie, D. M. 83
Baker, J. 160, 182
Barbour, I. G. 72, 87, 110, 115, 191
Barrow, J. D. 99
Barth, K. 95, 106, 136, 149, 209
Bartholomew, D. 114, 235
Barton, J. 149, 152–3
Bauckham, R. 209, 235
beauty 11–12, 24, 48–9, 93, 96, 98,
    141
Bennett, G. 49
Berger, P. 84
Bible 42, 56–7, 63, 136, 147–57, 200
big bang 71, 96, 104
biology 25, 28, 35, 84, 111

block universe 133–5
Bohm, D. 11, 30–1, 55, 76, 113, 122
Bohr, N. 63, 85, 90
Book of Common Prayer 50, 206
bottom-up interaction 116–17
bottom-up thinking 3, 26, 29–30,
    34, 159, 186, 196–7, 231–2
Bouyer, L. 205
Bowker, J. 35
brain 47–48, 53, 93, 111, 114, 116,
    214–16, 239
Brooke, J. H. 37, 235
Brown, W. 215, 235
Brümmer, V. 127
Buddhism 225–6, 231
Bultmann 149, 176, 178
Burrell, D. 90–1, 106

Cambridge University 1–2
Caird, G. B. 182, 235
Carnes, J. R. 22, 83, 235
Cartesian dualism 31, 52–3, 55
causal joint 74, 116, 117–19, 123
chaos 26–30, 101, 107, 112–13,
    117–18, 121, 122, 125, 141, 156–7,
    211, 233, 236, 238
chaotic systems 26–8, 112, 113,
    118–21
chesed 68, 210
Childs, B. S. 152
Church 50–1, 56–7, 59, 62, 65–6, 81,
    147, 154, 163–4, 166, 170–1, 173,
    178, 184, 190, 192–5, 198–204, 212
Church of England 2, 68
classical theology 69, 72, 74, 135, 190
Clayton, P. D. 48, 191
Coakley, S. 50

241

Cobb, J. B. Jr. 107, 115
communion 196, 218; *see also*
    Eucharist
Conway-Morris, S. 39, 44
Cosmic Christ 154, 221
creatio continua 36, 72, 108, 157, 212
creatio ex nihilo 36, 68, 72, 107–8,
    157, 221
creation 5, 32, 34, 36–7, 49, 51,
    59–60, 67–75, 85, 88, 91, 94–5,
    104–9, 114–16, 119, 122–9, 131,
    135–6, 138–42, 148, 155, 157,
    188–94, 204–5, 210, 212, 214, 217,
    219–24, 227, 231
Creator 3, 5, 32, 36–7, 41, 46, 49–51,
    53, 59, 61, 69–70, 72–5, 93, 96, 98,
    106–7, 123, 125, 128–9, 135–6,
    138–9, 141–3, 156–7, 189–93, 210,
    214, 217, 220–2, 227
cross 66, 70, 162, 177, 187, 206–7,
    211, 227
Cross, F. L. 190
crucifixion 161–2, 177–8
Cushing, J. T. 113

Darwin, C. 36–7, 43, 45, 56, 58–60,
    71, 98, 137
Davies, P. C. W. 29–30, 79, 98, 101,
    108, 120, 163–4
Dawkins, R. 102–3
death 38–41, 43, 46, 53, 57, 59, 62, 68,
    73, 105, 111, 116, 130, 136, 138–9,
    141, 150, 152, 157, 160, 166–7,
    170–1, 176–7, 182–3, 187, 189, 201,
    207, 210, 212, 214–21, 223, 226–7
de Broglie, L. 52
deistic 69, 74, 110, 117, 132, 156
deity 61, 69, 73, 88–93, 106, 155, 192,
    210, 229
demons 228
Dennett, D. 44
d'Espagnat, B. 19, 32–3
deterministic chaos 113, 121

Dirac, P. 23, 64, 96–7
dissonance 226, 228–30
divine nature 5, 61, 65, 69, 73, 90,
    124, 136, 195, 229
Dodd, C. H. 150, 161, 165, 175, 181,
    185
Donne, J. 212
dual-aspect monism 48–9, 111, 114,
    215
Dunn, J. D. G. 153, 173, 183–4
Dukas, H. 137

Easter 161, 170–1, 173, 179–81,
    186–7, 199
Eastern Orthodox Church 82, 218, 221
Eccles, J. C. 214
Eddington, A. S. 20–1
Einstein, A. 12–14, 22–3, 45, 52, 63,
    71, 85, 97, 137, 189
enlightenment 9, 226–7
epistemology 31, 44, 76, 85, 113, 117,
    121
eschatology 167, 209–24
Eucharist 197–208, 240
evil 3, 35, 40, 67, 70–1, 73, 103, 127,
    137, 143, 187, 191, 209, 211, 227
evolutionary 36–7, 41, 43–6, 71–4,
    84, 96, 102, 105, 137, 141, 212, 214,
    231
exclusivism 228–30

Farrer, A. M. 35, 116
Feyerabend, P. 14–16
Feynman, R. 64–5
foreknowledge 171
forgiveness 168, 202, 206–7, 211–12
free-process defence 73, 141–2
Frye, N. 86, 150, 154
fundamental physics 18, 20, 26, 61,
    92, 96–7, 102, 137
future 3, 21, 26, 29, 34, 38, 76, 112,
    116, 118, 119, 122, 124, 133–6, 139,
    171, 207, 209–14, 217, 222–4

Gage, P. 214–15, 219
Gandhi 227
genetic code 105
genetic mutations 43, 73, 98, 141
genome 43
Gifford Lectures 3, 89–90, 129, 159, 173, 197
Gleick, J. 27, 112
God-consciousness 40
Goodwin, B. 44
Gospels 126, 131, 151, 159–61, 163–5, 167, 170, 172–3, 175, 178–83, 185, 195, 199, 202, 226
Gray, A. 59, 71
Green, C. 106
Griffin, D. R. 54, 107, 115

Haldane, J. B. S. 197
Hart, T. 209
Harvey, A. E. 175
Hawking, S. W. 92, 98, 106
Hanson, N. R. 10
Hefner, P. J. 73
Heisenberg's uncertainty principle 22, 29, 76, 113
Hengel, M. 166
Herbert, N. 29
Hick, J. 93, 164, 213
Hiley, B. J. 76, 113, 122
Hiroshima 71
Hodgson, L. 173
Hoffmann, B. 137
holistic causality 117, 123
Holy Spirit 62, 193, 199, 208
hominid 38–39, 41, 44, 46–9, 139
homo sapiens 26, 37, 44, 60, 105
human nature 36–50, 86, 111, 114, 151, 214, 216
Hume, D. 17, 92, 98, 102
Huxley, T. H. 37, 58–9

incarnational theology 70
inclusivism 229–30

information-bearing pattern 42–3, 215–17, 219, 223
Isham, C. J. 133

James, W. 81–2, 197–8
Jammer, M. 189
Jantzen, G. 115
Jenkins, D. 187
Jeremias, J. 168, 172, 198
Jesus 63–4, 68–70, 95, 126–7, 130–1, 148–51, 155, 157, 159–88, 195, 198–204, 206–7, 210, 217, 218, 220, 223–4, 226, 228, 230
Jews 63, 90, 167–8, 193, 228
Job 79–80, 155, 215, 227
joy 49, 102, 199, 211–12
Judaism 148, 162, 167, 184, 227–8

Kant, I. 20–1, 33, 84, 98
Kauffman, S. 44
Kaufman, G. D. 114
Keats, J. 43
Kelvin, Lord 59
Kenny, A. 88
kenosis 69–70, 72–3, 124, 136, 142, 191, 193, 220
Kierkegaard, S. 173, 188
Kingsley, C. 36, 59, 71
Koran 148; *see also* Qur'an
Kuhn, T. 13–16, 22

LaCugna, C. 193–5
Landau, L. 108
language 12, 38, 40, 42, 48, 57–8, 63–4, 68–9, 72, 74–5, 82, 86, 89, 91, 93, 98, 103, 110, 115, 123, 128, 131, 137, 150–1, 166, 172, 176, 179, 190, 192–3, 207, 219, 226
Lapide, P. 185–7
Latin Church 82
Lewis, A. E. 136
Lewis, C. S. 128, 132
Leslie, J. 99–101

liberation 206, 211
Livingstone, E. A. 190
Lonergan, B. 86, 152
Lossky, V. 192
love 46, 68–76, 86, 88, 91, 142, 153,
    156, 167, 187, 189–91, 193,
    196, 210–12, 217, 219
Lucas, J. R. 127

McDonald, H. D. 127–8
McDonald, J. I. H. 182, 185
Mackay, A. L. 90
McMullin, E. 108–9
Macquarrie, J. 173–4, 176–7, 186, 188
Malony, H. N. 215
Mandelbrot, B. 28, 45, 47
Margenau, H. 20
Mascall, E. L. 129
mathematics 1, 39, 47–8, 55, 96, 98,
    231
Maximus the Confessor 192
Maxwell, C. 52, 59, 80, 151
Mitchell, B. 85
miracle 123, 126–32, 174–5, 186
Mohammed 227–8
monism 48–9, 88, 111, 215, 227
Montefiore, H. 101
Moltmann, J. 2, 32, 66, 70, 94, 107,
    187, 209, 211, 218, 220–1, 224
morals 39–40, 47–9, 70–1, 73, 75,
    138–40, 142, 154, 176, 202
Moule, C. F. D. 165, 167, 169–70, 173
Munz, P. 44
Murphy, N. 117, 133, 215
Murray, M. J. 142

natural theology 74, 94–103, 147, 155
nature of reality 29, 33, 51–5, 88, 97,
    133
nature of science 9–24
nature of theology 79–87
Neill, S. C. 159
neo-Platonism 192

new creation 68, 69, 116, 157, 188,
    205, 219–24
New Testament 63–5, 70, 128,
    149–50, 153–5, 159–60, 165–6,
    168–9, 171–3, 176, 178–80, 182,
    184–5, 188, 193, 211, 215, 219
Newton, I. 13–14, 22, 45, 52, 112, 122,
    137
Newton-Smith, W. H. 16, 18, 22
Nicaea 65, 195
Nicene Creed 3
Niebuhr, R. 40
Nietzsche, F. 50

O'Collins, G. 177, 182
Old Testament 150, 152, 155, 169,
    172
Omnes, R. 120
omniscience 124, 136
ontology 31, 76, 113, 117, 121
Oord, T. J. ix, 5
Origen 125, 154
original sin 40
Owen, R. 58

Pailin, D. A 88, 90
Pais, A. 97
Palamas, G. 192
panentheism 35, 190–2, 220
Pannenberg, W. 164–5, 179–80,
    186–8
panpsychism 54
Paradise 219, 223
Parmenides 133
Parsons, T. 93
Pascal, B. 82, 96
Peacocke, A. R. 25, 35, 72, 108,
    110–11, 116, 119, 190–1
Penrose, R. 216
perichoresis 196
persuasion 18–19, 69, 94, 115, 147,
    191
Peters, T. 49, 157

physical world 2, 16, 21, 25–35, 58,
    61, 65, 71, 82, 89, 92, 95, 98, 101–2,
    111, 114–15, 120, 124, 129, 148,
    155, 195, 231
Pickering, A. 15, 97
Planck, M. 30, 52, 63, 108
Planck's constant 30
pluralism 228–30
Polanyi, M. 12, 86
Pollard, W. G. 117
Pontifex, E. 183
Popper, K. 17–18, 33
positivism 19, 33
prayer 34, 50, 68, 79, 110, 114,
    126–32, 163, 168, 197–8, 200,
    202, 205–7, 211–12, 218
Prigogine, I. 27–8, 76, 118, 121–2
problem of evil 3, 71, 139–40, 142,
    187
process theology 54, 69, 73, 107, 115,
    136, 190–1, 213
providence 74, 110–25, 193
psychology 80, 91, 111, 231
psychosomatic unities 41, 111

quantum theory 10–12, 22, 29–31,
    33, 39, 45, 54–5, 62–6, 76, 85, 90,
    92, 96, 100, 108, 111, 113, 117–18,
    122, 231
Qur'an 227; *see also* Koran

Rae, A. 100
Rahner, K. 188, 194, 223, 229
Ramsey, I. 86
realism 19, 21–3, 62, 113, 194
reductionism 25, 41, 134, 182–3
resurrection 43, 53, 62, 69, 116, 124,
    130, 136, 150, 157, 161, 176–88,
    199, 206, 207, 216–24, 227
revelation 62, 90, 149, 155, 168, 174,
    179
Robinson, J. A. T. 160, 188
Rolston, H. 43

Royal Society 4
Ruse, M. 46
Russell, R. J. 117, 121, 123, 133

sacraments 147, 221
sacred 40, 49, 60, 153, 162, 189, 225,
    228–30
Sanders, E. P. 162–4, 167, 174, 177–8
Schillebeeckx, E. 176
Schleiermacher, F. 213
Schrödinger equation 29, 115
Schweitzer, A. 165
Schwöbel, C. 212
scientific method 9, 13, 22, 164
Scripture 57, 63, 66, 147–58, 177–8,
    197, 226
Searle, J. 216
selfishness 40
Sherwin-White, A. N. 160
sociobiology 46
Son of man 162, 169–70, 172
Soskice, J. M. 150–1
soul 41–3, 53, 210, 214–19, 223, 226
Spinoza, B. 189
Stannard, R. 109
Stengers, I. 27–8, 118
Stewart, I. M. 112
Stokes, G. 59, 129
strange attractor 27, 112, 118–19
Strauss, D. 177
Stump, E. 142
subatomic particles 25–6
suffering 66–7, 102, 129, 136, 138,
    142–3, 154, 201, 209, 221, 226–7
survival 43–6, 48, 96, 102
Swinburne, R. 128, 131, 197, 214

Tanner, K. 212–13
Temple, F. 36, 59, 71
Templeton Prize 4
Ten Commandments 80
theodicy 34, 70, 73, 75, 138–40, 191
theory of everything 26, 44, 49, 92

theosis 192, 221
Tillich, P. 213
Tipler, F. J. 99
top-down causality 116, 119, 121
Torah 148, 161, 168
Torrance, T. 85, 107
Tracy, D. 152, 197
Trinitarian/Trinity 65–6, 68, 70, 91, 142, 180, 189–96, 199
truth 1–3, 5, 9, 12–13, 16, 22–3, 32, 48, 55–9, 61–2, 75, 79, 81, 90, 97, 137–8, 140, 148–9, 152, 157, 172–3, 185–6, 198, 216, 227–8
two-slits experiment 64

Urban VIII 57–8

van Huysteen, W. 44
Vanstone, W. H. 70–1
verisimilitudinous 31, 61, 66, 195, 230

Vermes, G. 168–9, 185
Volf, M. 212–13, 217–18

Wallace, A. R. 71
Ward, K. 53, 91, 107, 110
Welker, M. 201, 204, 206, 212–13, 218
Weinberg, S. 26, 92
White, V. 110
Whitehead, A. N. 54, 69, 107
Whitehouse, J. C. 32
Wilberforce, S. 37, 58–9
Wiles, M. 114
Wilson, E. O. 44, 46
wisdom 57, 79–81, 155, 172, 176
world faiths 225–32
worship 2, 25, 40, 53, 65, 93, 96, 126, 198, 204, 212
Wright, N. T. 160
Wuketits, F. 44

zimzum 220

Made in the USA
Monee, IL
23 March 2021